THE VAMPIRE BAT

THE VAMPIRE BAT

A FIELD STUDY
IN BEHAVIOR
AND ECOLOGY
*
DENNIS C. TURNER

The Johns Hopkins University Press * Baltimore and London

The Johns Hopkins University Press, Baltimore, Maryland 21218
The Johns Hopkins University Press Ltd., London

Library of Congress Catalog Card Number 74-24396
ISBN 0-8018-1680-7

Library of Congress Cataloging in Publication data
will be found on the last printed page of this book.

For my parents and brother

CONTENTS

ACKNOWLEDGMENTS

This field study would have been impossible without the training, guidance, critical review, and moral support of my dissertation advisor, Edwin Gould, and the assistance and generous hospitality of Werner Hagnauer and family, owners of Hacienda La Pacifica in Cañas, Costa Rica. Victor Villalobos, head cowhand on La Pacifica, and his son Carlos were indispensable in collecting field data. The Organization for Tropical Studies, Inc., and especially Jorge Campabadal and Liliana Eschevierria, were of great assistance in arranging logistical support in Costa Rica.

Charles Southwick, Paul Heltne, and James Tonascia were particularly generous of their time in discussing findings and new problems. Numerous other professionals have provided stimulating discussion and assistance throughout all phases of this project. For this I am indebted to Drs. Paul A. Opler, Jack and Sandra Bradbury, Theodore Fleming, Ray Heithaus, Don E. Wilson, P. F. L. Boreham, Richard Vargas, Helmut Buechner, Gordon Orians, W. J. Hamilton III, Daniel Janzen, Doris Stone, Ari van Tienhoven, Thomas Schoener, Ronald and Barbara Carol, Kenneth Glander,

Wayne Van Devender, Albert Schaffer, G. Clay Mitchell, R. J. Burns, Uwe Schmidt, Arthur M. Greenhall, William A. Wimsatt, and K. Mueller.

I thank Eric Pianka and the University of Chicago Press for granting permission to use figure 11, which first appeared in *The American Naturalist*, copyright 1966, the University of Chicago Press.

Special thanks are due to Dean H. Heusser, Swiss Federal Institute of Technology, for providing office space and computer facilities in Zürich, and to Hans Kummer and Hans Burla, University of Zürich, for commenting on the resulting manuscript.

I thank Ethel Pulaski for maintaining order and communications while I was in the field and Susan Hobson for typing the manuscript. Lastly, I wish a special thanks to my wife, Heidi, and to my parents, who provided moral support and other assistance throughout my graduate education.

This research was conducted while I was on a National Institute of Mental Health predoctoral fellowship, No. F01-MH47,294, at The Johns Hopkins University. Financial support from the United States Public Health Service (5S01, RR05445-GRS-Johns Hopkins), the American Society of Mammalogists grants-in-aid program, the Society of the Sigma Xi grants-in-aid program, and National Institutes of Health grant No. 5R01,NS09579 (to Edwin Gould) is gratefully acknowledged.

THE VAMPIRE BAT

CHAPTER 1

INTRODUCTION

The mammalian order Chiroptera, comprising the bats, has long captured man's imagination and inspired mixed feelings of fear and respect. The pre-Columbian Indians of Central and South America modeled jade artifacts after bats; and the Aztec nation had as one of its idols the bat god Tzinacan (Villa-R. 1968). Probably no species of bats has aroused stronger fear responses in man than those aroused by the blood-eating vampire bats. Indeed, the general fear of bats today may be associated with the public's incomplete knowledge of vampire bats.

De Oviedo y Valdes (1950 [1526]) and Benzoni (1967 [1565]) were the first Europeans to mention the blood-eating bats of the New World. The latter's account (book 2, p. 160) is quite vivid: "There are many bats which bite people during the night; they are found all along this coast to the Gulf of Paria and in other areas, but in no other part are they as pestiferous as in this province [Neuvo Cartago, today Costa Rica]; they have gotten to me at several places along this coast and especially at Nombre de Dios, where while I was sleeping they bit the toes of my feet so delicately that I felt nothing, and in the morning I found the sheets and mattresses with so much blood that it seemed that I had suffered some great injury. . . ." Charles Darwin (1890) was the first scientist to observe the true vampire bat drawing blood. Research remained dormant until the early 1930s, when an outbreak of paralytic rabies in Trinidad killed thousands of cattle and eighty-nine humans. Pawan (1936) and Torres and Queiroz Lima (1936) found that the vampire bat was the principal vector of the deadly virus.

THE VAMPIRE BAT SPECIES: BASIC BIOLOGY

There are three genera of true, hematophagous vampire bats, each having one species; they are all members of the subfamily Desmodontinae and the family Phyllostomatidae (Koopman and Jones 1970). Exclusively New World species, vampires probably originated in tropical regions (Herschkovitz 1969). *Desmodus rotundus*, the species with which the present study as well as all others cited are primarily concerned, is generally common from Mexico through Argentina and Chile, while *Diaemus youngi* and *Dyphylla ecaudata* are generally rarer within the same range (Villa-R. 1968). On the basis of metabolic rates, McNab (1973) has related the present-day distribution of *Desmodus* to the 10° C minimal winter isotherm both in Mexico and in Argentina and Chile.

It would be of little value in this text to review current knowledge of the vampire's functional morphology, anatomy, and physiology. Instead, I will briefly introduce each of these topics with respect to how the vampire differs from other bat species or how it is particularly adapted for its peculiar feeding habits. We shall start with the facts that *Desmodus* is highly mobile both on the ground and in the air, must find its prey animals during the night, and feeds exclusively on the blood of various vertebrates. What adaptations does the vampire have for performing these activities?

Aside from the usual morphological adaptations for flight which all bats possess, the vampire has others which facilitate movement while either on the prey or on the ground. *Desmodus* is fully capable of walking, running, and hopping, in addition to flying. The thumb is elongated and well developed (see plate 9), and the hind legs are well set and relatively strong. The vampire is quick to react to disturbances, and the selective advantage of agility to an animal that can feed on prey almost ten thousand times its size is obvious.

Desmodus has a well-developed sense of smell and is capable of olfactory orientation (Mann 1960; Schmidt 1973; Schmidt and Greenhall 1971). It has large eyes and good visual acuity (Chase 1972) relative to other chiropterans. Its echolocation system makes use of low-intensity calls, best suited for the detection of large objects (Vincent 1963).

The superior incisors are razor-shaped and serve to remove a small (3 mm) piece of flesh from the prey (see Greenhall 1972a for a description of biting and feeding mechanisms). Several authors have reported the presence of anticoagulant or fibrinolytic activities in the saliva of *Desmodus* (Romaña 1939; Mann 1951; and DiSanto 1960); and more recently Hawkey (1966 and 1967) has described a potent plasminogen activator and a substance capable of inhibiting aggregation of human platelets.

In contrast to the simple mammalian gastrointestinal tract, the vampire has a T-shaped gastroesophageal-duodenal junction and an elongated tubular stomach. Mitchell and Tigner (1970) have shown that the ingested blood first

PLATE 1. The common vampire bat, *Desmodus rotundus*. Notice the characteristic flat nose and the relatively large eyes.

enters the intestine and then overflows into the stomach, the primary site of water absorption (Wimsatt and Guerriere 1962). The stomach and intestines are capable of extreme distension, so that a bat appears spherical and bloated after feeding (Ditmars and Greenhall 1935). Laboratory studies have shown that vampires will ingest an average of 15 to 16 ml of blood per day, or almost 40 percent of their fasting body weight (McFarland and Wimsatt 1969). The blood diet of the vampire bat is singularly high in protein and low in fats and carbohydrates. Because of this nitrogenous food, the vampire is forced to excrete highly concentrated urea with considerable dispatch and to live on a tight water budget (McFarland and Wimsatt 1969).

OTHER FIELD STUDIES

Ditmars and Greenhall (1935) and Dalquest (1955) were the first to present the natural history of vampire bats; but these early studies were descriptive, and the need for quantitative examination of vampire behavior and ecology was evident. In 1966 the World Health Organization Expert Committee on Rabies declared vampire-borne rabies to be the primary livestock disease problem of Latin America. The U.N. Food and Agriculture Organization and the U.S. Agency for International Development, in cooperation with the U.S. Department of the Interior, began large-scale programs to find vampire control techniques, which culminated in 1972 with the announcement that a chemical control agent had been successfully tested in Mexico (Thompson, Mitchell, and Burns 1972; Linhart, Crespo, and Mitchell 1972), and control programs are now being organized in various Latin nations. (I will discuss various control methods and recommendations for vampire control based on my findings in chapter 9.)

With a few notable exceptions, most field studies of vampire behavior and ecology in recent years have been spin-offs of these control programs and, as such, limited in scope. Schmidt and Greenhall (1972) reported on interactions between feeding vampires under natural and laboratory conditions. Crespo et al. (1961), Goodwin and Greenhall (1961), and Villa-R. (1966) published assorted ecological observations on the vampire. Schmidt, Greenhall, and Lopez-Forment (1970 and 1971) were soon to follow with findings on predation rates, host preferences, population densities, and foraging activities of Mexican vampires. L.-Forment, Schmidt, and Greenhall (1971) studied movement patterns in Mexican cave populations, while Crespo et al. (1972) noted an effect of moonlight on foraging flights. Greenhall, Schmidt, and Lopez-Forment (1969 and 1971) have published the only observations on the mode of attack by vampires, using night-vision telescopes to view bat behavior at the cow.

Of the non-control-oriented researchers, Wimsatt (1969) was first to report experimental results of a field study in Mexico. He stressed movement and activity patterns, as well as feeding efficiency. Young (1971) has conducted the only other long-term study of vampires at one study site (in Costa Rica), and his findings offer a basis for comparison with my results. A comparison between three Central American bat communities, one of which was the subject of the present study, was made by Fleming, Hooper, and Wilson (1972). It stressed community structure, reproductive cycles, and movement patterns of individual bat species, including the common vampire. Heithaus, Fleming, and Opler (in press) have continued studying resource utilization by bats in the same community.

THE PROBLEM

Several authors have suggested that vampire populations have soared since domestic animals were brought to the New World. Domestic animals afforded vampires a more accessible and more plentiful supply of blood than did the native wildlife (Dalquest 1955; Greenhall 1968 and 1972*b*; and Villa-R. 1968). From theoretical models MacArthur and Pianka (1966), MacArthur (1972), and Emlen (1966) have predicted that a more productive environment, that is, one in which food is common, should lead to increased selectivity in dietary habits and that an environment where food is scarce promotes indiscriminate feeding. Schoener (1971) has reviewed these models in an attempt to formulate a general theory of feeding strategies. If vampire populations have indeed soared with the increase in domestic stock, these models would predict a high level of selectivity among that stock if a basis for selectivity could be demonstrated.

In testing this central hypothesis, I attempted to emphasize the interaction between this mammalian predator (also an obligatory parasite, in the classical sense) and its prey from a comparative and longitudinal point of view. Although prey selection by *Desmodus* was the prime focus of my field research, a review of the literature revealed a general lack of quantitative information about related aspects of vampire behavior and ecology for any single study site. Thus, to obtain as complete a picture of the predator-prey interaction as possible, I examined (1) the vampire population size; (2) the hunting behavior of the vampire, including its activity patterns and the prey species', the location of the prey by the vampire, the hunting range and foraging time, and the success of the hunt; (3) prey selection by vampires and the basis for prey preferences; (4) the social behavior of the vampire; (5) seasonal trends in various aspects of the predator-prey interaction; and (6) differences between vampire populations.

In the following pages, I will describe the dynamic interaction between one predator, *Desmodus rotundus*, and its prey species. Although this description is not yet complete, hopefully the results of this field study will serve to stimulate future research and interest among vertebrate zoologists, ecologists, behaviorists, physiologists, public health specialists, and agriculturalists.

CHAPTER 2

STUDY SITES
AND GENERAL
METHODS

SELECTING THE
STUDY SITE

As stated in the introduction, *Desmodus* may be found as far north as Mexico and as far south as Argentina and Chile. I selected the Central American Republic of Costa Rica as the host country for my research for several reasons. A preliminary survey conducted while I participated in a field ecology course conducted by the Organization for Tropical Studies, Inc., confirmed the presence of vampires in various geographical locations within Costa Rica. Other vampire studies had been conducted in Mexico and Trinidad; thus, given the widespread species distribution, information on the behavior and ecology of the species in a different nation could be useful. Greenhall (1970*b*) recommended that future vampire studies be conducted in the

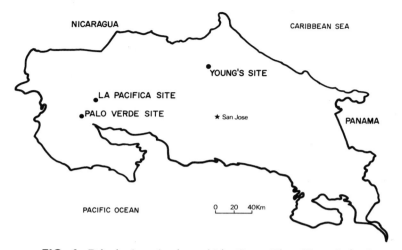

FIG. 1. Principal study sites within Costa Rica. The relative location of the site used by Young (1971) and San Jose, the capital, are also shown.

cattle-raising areas of Central America, and in particular the Nicoya region (Guanacaste Province) of Costa Rica. Young (1971) had published the only other study of Costa Rican vampire ecology, and this was undertaken in the Atlantic wet lowlands, where cattle are not generally abundant. Excellent logistical support was available from the Organization for Tropical Studies, the Costa Rican Ministry of Agriculture and Cattle, and the Ministry of Public Health, and the political and economic stability of this republic favored a long-term study by a North American.

I used five criteria in selecting the principal study site within Costa Rica: (1) the presence and relative abundance of cattle, and in particular various breeds of cattle; (2) the presence and relative abundance of *Desmodus rotundus*; (3) the presence of forest tracts which contained potential wild hosts; (4) the degree of cooperation that could be expected from ranch owners; and (5) the availability of living quarters for a long-term study.

In March 1972 I selected Hacienda La Pacifica, 4 km northwest of Cañas, in the province of Guanacaste, as the principal site for the fifteen-month investigation (see figure 1 for the location of study areas). La Pacifica is situated in the Dry Tropical Forest life zone (Holdridge 1967), where most of the annual rainfall of 1790 mm (a thirty-year average) comes between mid-May and mid-November (the extreme effects of the seasonal rains may be seen in plate 2). The year 1972–73 was an exceptionally dry year in Central America; still, most of the 934 mm of rain falling on La Pacifica came within the normal wet season. The mean annual temperature is 28° C (a seven-year average), with an average daily temperature variation of 9° C. Relative humidity averages about 75 percent. The climate of the area is governed by the trade-wind system; during the dry season, winds pick up noticeably. The deciduous forest

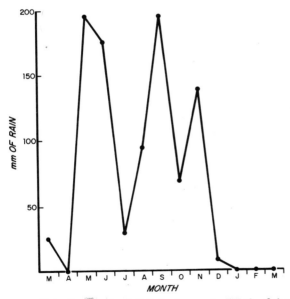

FIG. 2. Total rainfall during each month of the study period on La Pacifica. Data are from a federal weather station located on the ranch.

PLATE 2. The extreme effects of seasonal rains on the trees and pastures of La Pacifica. The upper photo was taken during the mid-rainy season, and the lower photo, during the mid-dry season.

loses most leafy matter during the dry season; only the gallery forests along rivers remain relatively unaffected.

How well did La Pacifica fit my criteria for a prime study site? First, the 1,300 hectare ranch had an average population of 1,187 head of cattle, consisting of Brahma (tropical Zebu) breed and Brown Swiss (European) breed individuals. The ranch also maintained about 35 horses, 10,000 laying hens (in non-vampire-proof cages), and a few other domestic species, which facilitated my host-selection studies. Second, Fleming, Hooper, and Wilson (1972) had conducted an extensive bat-community survey showing *Desmodus* to be fairly common on this ranch, yet not so abundant as to inhibit a study of the entire population. Third, two relatively large tracts of forest (125 ha and 190 ha) maintained on the property contained potential wild prey for the vampires. Deer, peccaries, monkeys, opossums, raccoons, and other wild animals had been seen on the ranch. Fourth, the ranch owner, Werner Hagnauer, was extremely cooperative in allowing experimental manipulations of his domestic stock. And last, because the ranch owns and operates its own restaurant and *cabinas*, excellent living facilities were available directly at the study area.

Given the importance of cattle herds in the vampire's daily life, a brief description of the ranch's stock management is in order. The herds on La Pacifica are carefully managed; accurate records of births, deaths, pregnancies, and herd movements are maintained in an index-card registry. Each animal is individually numbered, which facilitates recording data. Throughout my study much coordination between Hagnauer, Victor Villalobos, the head cow hand, and myself was necessary; this was accomplished with the use of a map showing herd locations and the card file in the ranch office.

Table 1 illustrates the relative stability of the cattle population and herd composition during the research

TABLE 1

Herd Composition on La Pacifica during the Study Period

	Sample date							
	5/28/72	7/2/72	9/24/72	10/31/72	12/12/72	1/28/73	3/27/73
Total bovines	1,306	1,284	1,208	1,108	1,191	1,123	1,091
Bovine types (in percent)							
Brahma	87.9	86.7	87.9	86.7	87.2	87.0	87.2
Brown Swiss	12.1	13.3	12.1	13.3	12.8	13.0	12.8
Heifers[a]	11.5	9.8	9.9	11.5	10.4	10.6	12.7
Pregnant/maternity[b]	25.0	18.1	16.6	18.0	22.9	16.5	27.1
Calves[c]	14.9	15.7	17.5	16.9	22.1	19.4	19.9
Cows to bulls[d]	18.2	27.1	24.4	20.6	16.3	29.4	14.3
Males[e]	29.2	28.2	30.4	31.8	27.5	23.2	27.6
Bulls	1.1	1.0	1.1	1.2	0.9	1.1	1.2

[a]Females eight to twenty-one months old.
[b]Cows pregnant for nine months and cows remaining with calves for eight months.
[c]Weaned at eight months.
[d]Breeding cows and heifers.
[e]Non-castrated males, eight to twenty-four months old.

period. However, for the five years prior to 1972–73, as more pastures were cut, the overall population steadily increased from 400 to 1,300 head. As mentioned before, there were two breeds of cattle: Brahma (*Bos indicus*) and Brown Swiss (*Bos taurus*). The only purebred Brown Swiss were the bulls; henceforth, when I refer to the Brown Swiss or the Swiss breed, I am referring to Brown Swiss-Brahma mixtures. The two breeds are maintained together in each of the ranch's herds. With the exception of herds with calves and breeding herds, males and females are maintained separately. Neither young nor adult males are castrated.

At present, the 1,300 ha ranch is divided into irrigated pastures separated by tree rows and by the two tracts of forest. It is bordered on the east by the Rio Corobici and on the west by the Rio Tenorio rivers, which join together 4 km

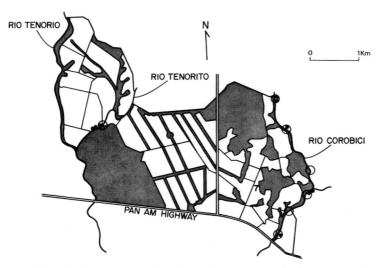

FIG. 3. Hacienda La Pacifica. Shaded areas indicate trees; dots mark known vampire roosts; and circles show the main netting sites used during the study. Pasture number 7 is indicated for later reference.

south of the ranch. The various herds are rotated from pasture to pasture as the fodder is depleted. Beef stock is usually grazed on *pangola* pastures, with a mean five days per rotation; breeding stock is grazed mainly on *jaragua* pastures, with a mean six days per rotation.

As my study on La Pacifica progressed, the value of comparative data from other study sites and vampire populations became more apparent. Two additional sites were occasionally visited (see figure 1). Finca Carrizal, 4 km south of Cañas, owned by Antonio Tac San Lam, maintained 700 head of Brahma stock and 27 horses on 280 ha. This ranch was selected for the absence of European breed stock in the immediate vicinity and for the presence of mixed-breed stock on surrounding ranches. Small vampire roosts were located in hollow trees at the center of this property. Between February and May 1973 I also made five visits to Cueva del Tigre, Hacienda Palo Verde, Comelco Property, also in the province of Guanacaste. Comparative data were gathered from this cave population of over 1,000 vampires. This site is also within the Dry Tropical Forest life zone, and rainfall patterns are similar to those on La Pacifica, 27 km to the northeast. This large ranch is believed to maintain around 5,000 head of cattle with little systematic management. For the most part, cattle are free to roam where they choose through forest, scrub vegetation, and open pasture.

GENERAL METHODS

Capturing and Processing the Bats. To collect the vampires, I used the standard technique of mist netting (Greenhall and Paradiso 1968), with 12 m long, 36 mm mesh, nylon mist nets stretched between two uprights. Bonaccorso and Turner (1971) had previously reported that *Desmodus* was predominately netted within 2.5 m of the ground in both heavy forest and open pastures. Usually from five to eight nets were used each netting night. To in-

PLATE 3. The author's assistant opening a mist net at sunset. Twelve-meter nets of nylon net mesh too fine to be detected by the bat's echolocation system were stretched between two uprights.

sure maximum capture rate, nets were set along or over rivers, streams, or pathways. Nets were rarely set in open fields, where the vampires fed, because preliminary efforts at catching bats in open pastures yielded few, if any, bats. Although the net sites shown in figure 3 were regularly utilized each month, between successive netting nights the net configuration within a site was changed to minimize the chances for the vampires to learn the locations of the nets.

Before opening the nets at dusk, the net site, date, number of nets, moon phase, and weather conditions (precipitation, cloud cover, wind conditions) were recorded.

PLATE 4. The author removing a bat from a mist net. Gloves were always worn as protection against bites. A battery-operated headlight is standard equipment for batting.

Once each hour the nets were cleared of bats, leaves, and other animals. I used two gloved hands to remove vampires, and an assistant recorded the following data on a standard form: (1) time of capture, (2) net number, (3) sex and reproductive condition, (4) forearm length, (5) number of ectoparasites, (6) stomach condition, (7) weight, (8) band number, and (9) special comments.

For time of capture, the hour immediately preceding removal from the net was recorded. Each net within the net site was numbered, and a map was drawn for later reference. Vampires are easily sexed (Greenhall 1965). The reproductive condition of females was determined by external palpation and teat manipulation after Fleming, Hooper, and Wilson (1972). Certainly embryos less than a month old were missed by this method; in a behavioral study of individuals in a limited population, it is sometimes necessary to sacrifice accuracy rather than specimens. Forearm length was measured with a standard millimeter ruler. When a vampire has fed, its stomach distends to such an extent that the bat appears spherical (Ditmars and Greenhall 1935). For recording stomach condition, I used the codes "1" for bats which had obviously not fed, "3" for full bats, and "2" for bats whose feeding activities I was unable to determine. I weighed the bats to the nearest gram, using a Pesola scale (100 gm) and weighing bag. With the exception of bats sacrificed for other purposes, I banded each specimen with a U.S. Fish and Wildlife numbered metal bat band around the forearm before releasing it—males on the right forearm; females on the left—which facilitated recording data on known individuals throughout the study period. Lastly, I recorded additional comments, such as coat color or presence of wounds.

Stomach Content and Rabies Analyses. To determine the frequency of utilization of wild versus domestic prey by vampires, I dissected blood meals from the stomach, intestines, and colon. One smear of each blood meal was placed on a circular filter paper (Whatman no. 1), labeled, dried, and air-mailed to the Immunology Laboratory at the Imperial College Field Station in Berkshire, England. There, precipitin and haemaglutination tests using anti-sera from human, domestic cow, horse, pig and fowl, wild deer, monkey (Howler), and rodent blood were performed. Greenhall (1970*a*) described how to interpret the results of these tests for vampires. Three different feedings can be determined by using blood smears from different regions of the digestive tract. I purposely kept sample size small to minimize the effect of lowering the vampire population on other concurrent studies. I used these results to compare the wet-dry season, full-new moon, and male-female feeding patterns of wild versus domestic prey utilization.

After dissecting the blood meals from these bats, the carcasses were deep-frozen for several days on the ranch; they were then packed in ice and hand-carried to Dr. R. Vargas at the Virus Laboratory of the Ministry of Public Health in San Jose, where tests for the presence of rabies virus were conducted. Standard fluorescent antibody (FA) tests were performed on brain and brown fat tissues; if results were questionable, Vargas continued with suckling mouse inoculations and further conducted FA tests on the suckling mouse brain tissues. Aside from the rabies status of vampires which had bitten me, I was interested in determining any seasonal trends within the population.

Herd Bite Surveys. From the precipitin and haemaglutination tests on La Pacifica specimens and from Greenhall's (1970a) results in Trinidad and Mexico, it became quite clear that vampires relied heavily on domestic stock for blood meals. To obtain information on predation rate and prey selection within the domestic stock types, each of the herds was surveyed daily for vampire bites. On a ranch as large as La Pacifica, it was impossible to conduct these surveys alone. Thus the ranch's four cowboys, one full-time assistant, and I shared this work. We surveyed the herds from horseback by moving in and around each animal as it grazed. From fifteen to twenty minutes were required to check a herd of one hundred animals. Usually surveys were conducted in the early morning before the blood drippings of a wound dried and chipped off the animal's coat. We recorded color of the bitten animal, brand number, and loca-

PLATE 5. Moving a herd into the corral for an intensive vampire-bite search.

tion of the bite on the body. On three occasions over the fifteen-month study, I checked the accuracy of the bite reports made by others, by spot surveying different herds all over the ranch, and found that only one bitten animal had been missed. This was not surprising, since a vampire wound is unmistakable (Dalquest 1955). Greenhall, Schmidt, and Lopez-Forment (1971) and Crespo, Burns, and Linhart (1971) have shown that vampires occasionally bite domestic cattle in different places on the body. To be certain that we were not missing bites because of their position on the body, such as in the armpit or around the hoof, whenever a herd was corralled and run through a cattle chute for an experiment, I carefully checked each animal for such bites, and none were ever spotted. Thus I am convinced that our bite reports were accurate. The horses were not surveyed as systematically, since their whereabouts were not always known. Horses are ridden by the ranch hands as well as by tourists, and everyone participated in reporting vampire wounds on the horses they rode.

Observation of Bats and Cattle. I confined my direct observations on vampire social behavior to the Palo Verde cave population. There it was relatively easy to observe the bats with a red-cellophane-covered head lamp, which rarely disturbed the bats. The locations of subgroups within the large limestone outcropping were also mapped. However, with each succeeding visit to the cave, more roosting crevices were noted; therefore, these data should not be regarded as complete. Additionally, vampires were frequently observed flying over and around a cattle herd.

I suspected that the nocturnal behavior of the different prey types might influence vampire hunting behavior and prey selection. Specifically, I was interested in behavior patterns that might increase or decrease the prey's exposure to potential vampire attacks. Studying bovine behavior at night presented a problem, namely, viewing the animal or herd. To solve this, I followed two procedures. From June until mid-August 1972, I used a U.S. Army Night Vision Starlight Scope, on loan from the Smithsonian Institution, a portable device, weighing 2.5 kg, which intensifies the existing light of stars or moon 45,000 times. Magnification is about 4 ×, field of view is 10°, and focus is up to 400 m. I made all herd observations from a position low to the ground at a stationary point in a pasture. My silent presence usually did not disturb the herd; if it did, observations on that night were discontinued. The second procedure, used to locate and observe the herd for short periods, involved the use of a battery-operated headlight while on horseback. Twice this caused a stampede; however, the herds quickly adjusted to my presence with the light. Even though I could see the herds when the moon was up, I continued to use the headlight to avoid any effect a change in my behavior might have on the herd's behavior. Data on activity patterns, distance to nearest neighbors, specific position within a resting herd, and resting locations within a pasture were gathered in this manner.

CHAPTER 3

POPULATION ESTIMATES

A reasonable estimate of the vampire population size is essential as a basis for interpreting my results. Thus I will begin with a chapter on population estimates, even though (1) some of the numerical values used in making these estimates are data reported later in the text, and (2) I did not have such an estimate, myself, until halfway through the field study. From Fleming, Hooper, and Wilson (1972) and from preliminary observations, I did know that a vampire population existed on La Pacifica; from casual observations on the frequency of livestock bites by Hagnauer (personal communication), I knew that the number of vampires was not so large as to inhibit a study of the entire population.

I used three methods to estimate the number of vampires on La Pacifica. Two of these are based on mark-recapture information; the third is inferential and is based specifically on vampire feeding habits. For the mark-recapture techniques, described by Schumacher and Eschmeyer (1943) and Jolly (1965), I treated each monthly sample as one point in time. Recaptures within the same month were excluded from the analysis.

The calculation formula of Schumacher and Eschmeyer is

$$N = \frac{\Sigma C_t \, m_t^2}{\Sigma m_t \, R_t} \, ,$$

where C_t is the number of individuals caught during a time interval, t; m_t is the number of marked individuals at large during t; and R_t is the number of recaptures during t. The numerical data, which appear in table 2, result in an estimate of 588 individuals (95 percent $C.L.$ = 506 and 702).

Jolly (1965) gives the formula for calculating the marked population size at time interval, i, as

$$M_i = r_i + \left(\frac{Z_i}{R_i} \right) \, a_i,$$

where M_i is the estimated marked population; r_i is the number of recaptures during time interval, i; Z_i is the total number of individuals marked before i, not caught during i, and then caught later; R_i is the total number of individuals marked during i and later recaptured; and a_i is the total number of individuals marked during i. I calculated an M_i for each monthly sample; then these values had to be divided by the proportion of animals marked in each sample to give a total population estimate for each month (the numerical data appear in table 2. The mean total population size calculated in this manner was 632.7 bats (SD = 482.7).

Obviously, the above methods yielded population estimates with such extreme variability that I could say nothing about the actual population size. Knowing in advance of inherent problems with these methods, I developed a third

TABLE 2
Data Used in Calculating the La Pacifica Population Estimates

t	Schumacher-Eschmeyer Method			Jolly Method		
	m_t	C_t	R_t	M_i	Proportion of Animals Marked	Estimate of Population Size
March	0	36	0	—	—	—
April	36	49	5	34	0.10	340
May	80	15	1	—	—	—
June	94	31	1	37	0.03	1,233
July	124	30	6	123	0.20	615
August	148	24	9	109	0.38	287
September	163	26	6	135	0.23	587
October	183	28	12	184	0.43	428
November	199	33	14	185	0.42	440
December	218	36	8	114	0.22	518
January	246	39	17	809	0.44	1,839
February	268	40	19	229	0.48	477
March	289	38	18	98	0.50	196

method during the course of my field work. It is inferential and takes into account (*a*) the observed number of cattle bitten per night, (*b*) the number of vampires potentially feeding from the same wound through the night, (*c*) the number of days between feedings, and (*d*) the number of times per night a foraging vampire feeds. The calculation formula is

$$N = \frac{abc}{d}.$$

I assume the following values for each of the above factors. Let *a* equal 2.8, or the number of bitten animals per night (the average for 14 months) on La Pacifica (see chapter 7). Let *b* equal 1, 10.5, or 21 bats feeding from the same wound through the night. Greenhall, Schmidt, and Lopez-Forment (1971) had reported the maximum number of bats ever seen feeding from the same wound as 7 over a 3-hour period. Thus in a 9-hour foraging night (from 1900 to 0400 hours, allowing search and return time) 21 bats could possibly use the same wound. The estimate of 10.5 bats per wound takes into account the foraging time available during each moon phase (see chapter 4) and the observation of 7 bats per 3 hours. Let c equal 3 days between feedings, or the maximum tolerated before starvation (Greenhall 1965 and 1970*a*). I chose the maximum value of this factor, given the extremely low number of bitten stock animals per night on La Pacifica, compared with data from other vampire studies. And let d equal one feeding per foraging vampire. Dalquest (1955), Wimsatt (1969), Young (1971), and I (see chapter 4) have found that vampires forage only once nightly. Then N (minimum) equals (2.8)(1)(3)/(1), or 8.4 bats; N (for available foraging time) equals (2.8)(10.5)(3)/(1), or 88 bats; and N (maximum) equals (2.8)(21)(3)/(1), or 176 bats.

The estimates resulting from the Schumacher-Eschmeyer (1943) and the Jolly (1965) formulae are undoubtedly too high. Given the bite rate actually observed on La Pacifica and the maximum number of bats that could possibly feed from the same wound through the night, these estimates are at least twice as large as expected. On the basis of several roost surveys I conducted, the estimates from my method are within reason. On the Rio Corobici river there were a mean 46 vampires (the range was 34–57) inhabiting tree roosts, and since this is one of two rivers bordering the ranch, I suspect a resident population of 100–150 bats. This suggests that the actual b and c values which I chose are close to the observed maxima. Schmidt, Greenhall, and Lopez-Forment (1970) used a similar approach, with the factors of my formula, to produce a population estimate for their Mexican study site. La Pacifica covers about 1,300 ha of land; an estimate of 100–150 vampires thus yields an average density of about one vampire per 9–13 hectares, or one vampire per 8–12 stock animals.

We now have La Pacifica estimates ranging from 588 and 633 bats to 100 and 150 bats. Why such a large discrepancy? Let me eliminate two criticisms of the numerical values I chose for the inferential formula. First, as I have already stated in chapter 2, I examined the accuracy of bite reports and found that we were not missing bites because of the distant location of a herd or because of a hidden wound. G. Clay Mitchell, of the A.I.D. Vampire Control Project, was so astounded at this low bite rate that he personally checked other ranches in the Guanacaste area, and he found similar results. Second, low bite rates were not a result of La Pacifica vampires feeding either on other ranches in the vicinity or on wild prey. The stomach content analysis revealed exclusive use of domestic stock (see chapter 5). Given the preferences among domestic stock types which I established (see again chapter 5), it is unlikely that the La Pacifica vampires flew to neighboring ranches to feed. I must conclude that the bite rate on La Pacifica is typical of that on other ranches.

I banded a total of 279 vampires during the 15-month study on La Pacifica and had an average recapture rate on the order of 28.3 percent over that period. This is not significantly different from the 20.2 percent reported by Fleming, Hooper, and Wilson (1972) for La Pacifica vampires in 1970–71 ($\chi_1^2 = 0.37$, $P > 0.5$; 23 recaptures/114 banded vampires), and it is similar to the 25 percent recapture rate in an open Mexican field found by Schmidt, Greenhall, and Lopez-Forment (1971). Yet if I estimate the vampire population at 100 bats and banded 279, what happened to the other 179? I do not think it is safe to assume a mortality rate of over 60 percent, since banding records on adult bats do not include infant mortality. Nor do vampires chew and bite at the metal armbands, potentially dis-

PLATE 6. A typical netting site along a river flyway. Bats fly from one area to another along rivers, roads, and other cleared pathways. Bat capture rate along such a flyway is maximal.

lodging the markers, as other species do. Instead, I postulate two reasons for the discrepancies. First, I caught most bats along rivers and pathways; thus many bats may have been transients, using the river as a flyway to other foraging grounds. My results on roost-group turnover support this interpretation (see chapter 6). Second, I suggest that there are really two populations in the area, one resident and one made up of transients, and that the residents may aggressively exclude the transients from feeding on their cattle. I will elaborate on this hypothesis and offer supporting data in chapter 6. The one-time capture of these transient vampires would have biased the results of the Schumacher and Eschmeyer (1943) and Jolly (1965) calculations upwards.

Only one other field study has produced an estimate of vampire population size, and the data are not comparable, since the authors gave no density estimate or foraging range size (Schmidt, Greenhall, and Lopez-Forment 1970). However, predation rates (number of bites per animal per night) have been published for other sites, and I will discuss these in chapter 7. For the size of the resident vampire population, the predation rate on La Pacifica was extremely low. That many vampires feed from the same wound throughout the foraging night is one reason for this. There must be some selective advantage to successive utilization of the same feeding wound, and Greenhall (1972a) may have provided the clue. He found that selecting the biting sites on cattle can take as long as forty minutes if the animal does not already have a wound and as little as just a few minutes when a bat is reopening a wound. Thus, since it takes less time and energy to feed from a wound that has already been made, the bats are less exposed to other predators.

Other factors might also be related to the relatively low number of bitten animals on La Pacifica. I have already mentioned that for the five years prior to this study, the cattle population increased from four hundred to twelve hundred head. Perhaps as a response, the vampire population is increasing. However, the maximum monthly proportion of pregnant and/or lactating female bats which I observed does not support this (see chapter 7). An increase in the vampire population would also imply that the vampire population was limited by food just a few years before, which is extremely doubtful. Costa Rica experienced a rabies epidemic in 1968 (Centro Panamericano de Zoonosis 1972), and there might have been a decrease in the number of vampires as a result. But this would still not explain the low number of bitten livestock relative to the population size four to five years later. I conclude that the major factor is multiple use of few wound sites, and we begin with a population of 100 to 150 vampires utilizing 1,200 head of domestic stock over 1,300 hectares.

CHAPTER 4

HUNTING
BEHAVIOR
OF THE COMMON
VAMPIRE

In this chapter I will present the results of experiments and observations on four aspects of vampire hunting behavior: (1) the relationship between activity patterns of the predator and prey species; (2) location of prey animals by the vampire; (3) hunting range and foraging time; and (4) success of the hunt.

THE RELATIONSHIP BETWEEN
ACTIVITY PATTERNS OF THE PREDATOR
AND PREY SPECIES

Schoener (1971) considers optimal foraging period as one of four key aspects in animal feeding strategies. Although no formal theory yet predicts the optimal placement of feeding periods over activity cycle, they will certainly occur (a) when availability of food is high and (b) when

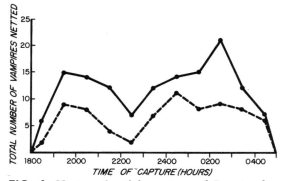

FIG. 4. Nocturnal activity pattern of *D. rotundus* during the new-moon period. The solid line represents both males and females; the broken line is for females only. Data are pooled from twelve new-moon nights.

predator pressure is low. Jahoda (1973) suggested that both of these factors were selective pressures acting on the lunar periodicity of grasshopper mouse foraging. The emergence times of captive genets, kinkajous, and rangtails have been examined by Kavanau and Ramos (1972), who interpreted the cycles as adaptative for predation on smaller mammals and for the avoidance of falling prey to larger animals.

Although *Desmodus*, like most other bats, flies during the dark of night, I assessed its nocturnal activity patterns in relation to both the lunar cycle and to the nocturnal activity patterns of its principle prey species, domestic cattle (see chapter 5). To obtain the basic activity patterns in relation to the lunar cycle, I plotted the number of vampires captured during each hour of the night under new-moon, first-quarter, and last-quarter moon phases. Figures 4 and 5 include data from twelve, six, and five netting nights, respectively; these show data only from nights on which "all-night netting" was conducted. The curve for twelve new-moon nights is bimodal; the curves for six first-quarter and five last-quarter moon phases are distinctly unimodal, each peaking when the moon was below the horizon. These patterns are not the result of vampires seeing the capture nets when the moon is up. Observations at the La Pacifica and Palo Verde roosts indicate that most vampires do not fly during these periods. To further prove this, I compared the number of bites on an equal number of animals and animal types during the four days around calendar new moon and the four days around calendar full moon for three lunar cycles during the wet season. Table 3 illustrates the reduction in number of bitten animals during the full-moon phase as opposed to the number bitten during the new-moon phase. Likewise, this reduction in bitten livestock is not due to the selection of wild prey during the full-moon period (see chapter 5).

But the moon may not be the only environmental factor affecting the nocturnal activity of vampire bats. Prey

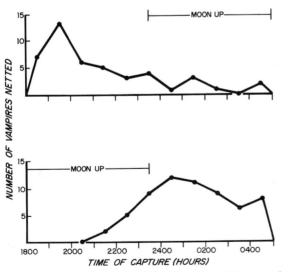

FIG. 5. Nocturnal activity patterns of *D. rotundus* in relation to the lunar cycle. The upper curve is from five netting nights when the moon rose at 2330 hours; the lower curve is from six netting nights when the moon set at 2330 hours.

TABLE 3

Reduction in Number of Bitten Animals during Three Full-Moon Periods

Total number of bitten animals		Percent reduction[a]
New moon	Full moon	
20	1	95
26	10	62
49	4	92

Note: Data were taken during the wet season months between May and October. One lunar period is defined as the four days around the appropriate calendar date.

[a]Mean percent reduction, 83; *SD* = 14.9.

availability or accessibility may change with lunar conditions and directly or indirectly affect vampire foraging behavior. Payne, Laing, and Raivoka (1951) and Lampkin and Quarterman (1962) have reported that both European and Zebu cattle graze at night in the tropics. However, neither study took position or phase of the moon into account. To determine the existence of such an effect, I observed the same herd on seven first-quarter and six last-quarter moon nights. For each hourly interval between 1800 and 0500 hours, I recorded whether more than 50 percent of the herd was grazing (standing) or was bedded down. Although cattle do not sleep, that is, do not lose consciousness, like other species (Brownlee 1950; Balch 1955), for convenience I will occasionally refer to a sleeping herd. Figure 6 shows the nocturnal grazing and sleeping activity of La Pacifica cattle in relation to the moon. The data strongly suggest peak grazing periods when the moon is up and sleeping periods when the moon is down. To confirm the time when vampire wounds were actually inflicted, I surveyed between three and five herds for bites on twelve nights before and after the grazing period and/or before and after the sleeping period. Eighty-eight percent of the bites ($n = 51$) were made when the herds were bedded down and not standing or grazing.

Brown (1968) reported that *Desmodus rotundus* is active during the darkest part of the night, but he failed to mention the bimodality of his activity curve. Certainly his sample size (13) was too small for a generalization; however, my curve for 135 vampires netted on 12 new-moon nights is very similar to his. At this time I can only speculate on reasons for the bimodality of activity during this lunar phase. This pattern would result if vampires fed twice each night, once during the early evening and once again toward morning. Only one vampire known to have

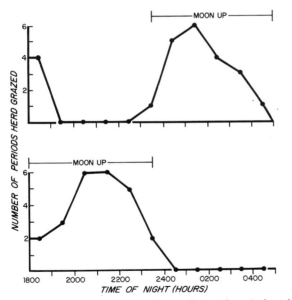

FIG. 6. Nocturnal grazing activity of cattle in relation to the lunar cycle. The number of periods the herd grazed refers to the number of one-hour intervals in which over 50 percent of the animals grazed. The upper curve is based on six observation nights; the lower curve on seven nights.

fed twice the same night was caught during the entire study. Dalquest (1955), Wimsatt (1969), and Young (1971) also found that vampires foraged only once nightly. A bimodal curve might also result if vampires foraged early and then flew to a temporary nocturnal roost before returning to the diurnal roost, causing the second peak in activity toward morning. Vampires did not do this in Wimsatt's or Young's study areas. However, I cannot rule this out as a possible explanation for La Pacifica. One notes that if the two activity curves for first-quarter and last-quarter moon phases are superimposed one upon the other, a bimodal distribution appears. Thus, the evening and morning peaks in vampire activity could result from two groups of vampires which begin feeding flights at different times, as Wimsatt (1969) suggested. Greenhall, Schmidt, and Lopez-Forment (1971) reported that vampires consistently arrived at the prey in groups of two to six. However, using Lloyd's (1967) mean-crowding method to analyze patterns in time, Heithaus, Fleming, and Opler (pers. comm.) concluded that *Desmodus* on La Pacifica foraged singly rather than in groups. This was my impression from roost-emergence patterns. Since most of my netting sites were along rivers and not at the feeding site, it is possible that this bimodality merely reflects the capture of bats going to and returning from the feeding site. Thus the dip in the curve of figure 4 could indicate time of feeding at the prey. I doubt this for several reasons. First, during other moon phases, peaks in vampire capture rates correlate with the times bites occur. Second, the amount of time required to secure a blood meal (estimates range from 0.5 to 3.0 hours) is too short to account for the bimodal spread (Dalquest 1955; Wimsatt 1969; Young 1971; and Greenhall 1972*a*).

Crespo et al. (1961) reported that vampires left the roost when darkness was complete and that they took preliminary flights to check the moonlight. Although a relationship between vampire foraging behavior and moon-light had already been demonstrated (Crespo et al. 1972), my study is the first to show an inverse relationship between vampire foraging patterns and cattle grazing periods, as well as a positive relationship between moonlight and cattle grazing periods. But relationships between two parameters are not necessarily causal. For instance, the lack of extra-roost activity by vampires during moonlit periods may be a predator-avoidance behavior. Predatory birds are known to chase and capture bats (Greenhall 1970b), and Schmidt found the skull of a vampire in an owl pellet. Nevertheless, when I compared the degree of reduced foraging activity during moonlit periods among vampires and other bat species, the vampire was most strongly affected (Turner, in preparation). To me, this suggests that predator avoidance is of less importance to the vampire than prey activity patterns. I conclude, then, that the vampire restricts its foraging to moonless periods, and this pattern is probably more strongly reinforced by the activities of the prey than by predator pressure.

LOCATION OF PREY ANIMALS BY THE VAMPIRE

The search for prey is certainly one of the key aspects in any predator's foraging strategy. We now know when the vampire forages; but information on how prey animals are found and where the vampire searches is also important in our general picture of hunting behavior.

As mentioned earlier, the cattle herds on La Pacifica are moved from pasture to pasture as the fodder is depleted. But they remain on each pasture for five or six days. Perhaps the vampire could learn the specific location of a herd and return to it for later feedings. If this occurs, I would expect the number of bitten animals to increase as more bats find the herd.

I recorded the number of bites on the first, second,

and last night a herd was in a pasture. Table 4 lists the number of bitten animals on each of these nights for nineteen consecutive pasture rotations of the same herd. Signed rank tests comparing bites on the first and second nights and on the first and last nights in a pasture revealed no significant differences ($T = 40$, $P > 0.1$; $T = 54.4$, $P > 0.1$). I also looked for trends over each set of days the herd was in a different pasture; none was found. Thus it *appears* from this experiment that vampires do not learn a herd's specific location. One factor, however, could lead to erroneous interpretation of these results. Several to many vampires could feed from the same wound each night, and the number of vampires feeding from each wound (not the number of wounds) could increase over time. Multiple feedings from the same wound have already been recorded by Greenhall (1972*a*) and Schmidt and Manske (1973) and were assumed in my population-estimating technique.

Wimsatt (1969), Villa-R. (1966), L.-Forment, Schmidt, and Greenhall (1971) and R. Burns (personal communica-

TABLE 4

Number of Bitten Animals on the First, Second, and Last Nights a Herd Was in a Pasture

First night	Second night	Last night	First night	Second night	Last night
0	0	5	2	6	1
1	2	2	3	0	0
3	2	2	6	3	0
3	3	2	4	0	0
1	0	0	1	0	0
2	1	1	0	0	1
0	5	6	4	0	2
3	5	5	3	0	0
4	1	0	4	0	2
5	4	2			

Note: Data are from nineteen consecutive pasture rotations.

PLATE 7. A typical hollow-tree roost used by the vampires on La Pacifica. The bats shift from roost to roost to maintain close proximity to preferred herds.

tion) have reported that vampires tend to remain within an area and regularly visit the same roosts but, nevertheless, may shift roosts frequently. I noted that at least some vampires changed tree roosts along the Rio Corobici river fairly often, and I decided to test whether roosting location was related to the location of potential prey. If, in fact, more and more vampires move into a particular roost in response to prey location, the frequency of multiple feedings from the same wound or wounded animals could increase. I surveyed each of the known vampire roosts along the Corobici and recorded the number of individuals on six occasions when I knew the pasture location of the heifers, a preferred herd. In two places along the river, there were two vampire roosts within 5 m of each other (see figure 3); these were counted as one roost when tallying the number of vampires. I measured the distance from the pasture containing the preferred herd (the pasture varied) to each of the roost locations on a map.

Figure 7 is a plot of the number of vampires in known roosts along the Corobici as a function of distance from those roosts (in standard units) to the pasture containing the preferred herd. A Spearman Rank-Difference correlation test on these data revealed a significant negative correlation ($r_d = -0.53$, $P < 0.05$); the greater the distance from a preferred herd, the fewer the number of vampires in roosts. Data points circled are from surveys conducted when the preferred herd was located in pasture number seven. No clear-cut relationship exists between distance from the roosts to this pasture and the number of vampires in those roosts ($r_d = -0.11$, $P > 0.1$); pasture 7 is approximately equidistant from all of the vampire roosts (see figure 3).

Wimsatt (1969) and Villa-R. (1966) have suggested that roost shifts may be made on an opportunistic basis, related to prey distribution. McFarland and Wimsatt (1969) have shown that *Desmodus* exists on a tight water budget, and prolonging foraging flights to return to the original roost, as opposed to a closer one, would be energetically expensive. My findings that vampires maintain close proximity to preferred herds could support the latter statement. Not only would the bats benefit energetically by

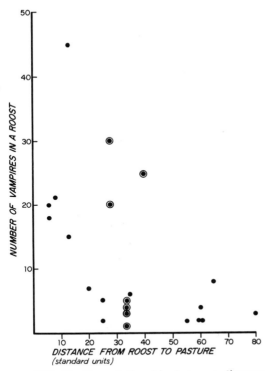

FIG. 7. The relationship between distance from a vampire roost to the pasture containing a preferred herd and the number of vampires in that roost. The pasture containing the preferred herd varied. Distances are in standard units and were taken from a map (1 standard unit = 33.5 m). Circled points indicate data taken when the preferred herd was in pasture number 7, approximately equidistant from all roosts.

flying to the nearest roost after feeding; that nearest roost would most likely be closest to the preferred herd on the succeeding foraging flight and thereby further reduce energy requirements.

I still doubt that vampires actually learn the specific location of a preferred herd. Rather, they probably have good spatial memory, as suggested for bats in general by Suthers (1970), and simply return to the nearest roost after feeding. My observations on several occasions when a preferred herd was moved to a new pasture indicate that the roost population increases dramatically the next night.

Additionally, it may be more difficult for vampires to learn the specific location of a cattle herd if that herd regularly changes its sleeping sites within the pasture. Looking at the first sleeping site and then the next successive sleeping site for twenty-four pairs, I found that on six occasions the herd returned to the previous sleeping location. A chi-square test revealed a significant difference ($\chi_1^2 = 5.04$, $P < 0.05$); a herd regularly changes its sleeping sites within a pasture. Could this be a response to vampire attacks? To answer this question I recorded whether or not a herd changed sleeping sites and whether or not animals were bitten at the previous sleeping site. A chi-square test for independence performed on the data in table 5 showed that bites occurring at the previous sleeping site had no effect on the choice of the next sleeping site ($\chi_1^2 = 0.04$, $P > 0.8$).

That a herd regularly changes its sleeping sites within a pasture may be a generalized predator-avoidance behavior. When a herd first rises in the morning, most constituents defecate, and feces are thus concentrated within the pasture (Hafez, Schein, and Ewbank 1969); the resulting odor might be an excellent cue for large predators. That vampire attacks at the previous sleeping site had no

TABLE 5
Relationship between Vampire Attacks and Choice of Next Sleeping Site by a Herd

Next sleeping site	Herd attacked	Herd not attacked
Identical	4	4
Different	8	7

effect on the choice of the next sleeping site came as no great surprise to me, given the low bite rate on La Pacifica. Hafez, Schein, and Ewbank described the leader-follower roles in a grazing herd that moved as a unit. The rare vampire bite would have to be inflicted on a leader animal before I would expect vampire attacks to affect choice of the next sleeping site; though undetermined, the probability of this was certainly low.

Little is known about the sensory basis of food-finding by non-insectivorous bats (Suthers 1970). I conducted two experimental attempts at assessing the use of vision and/or olfaction by the vampire to locate prey. If vision is used, one might expect a difference in color (white versus another color) of animals selected depending on lunar conditions. Table 6 shows the color of animals bitten during the full-moon and new-moon phases. I assumed that the proportions of white and nonwhite bovines remained the same over the year. A chi-square test for independence revealed no significant difference in color of animals bitten between the two lunar conditions ($\chi_1^2 = 0.11$, $P > 0.5$).

These experimental results are difficult to interpret because of improper experimental design. At this point, I am unable to conclude the vampire's use of vision to locate prey. Although vampires fed on prey irrespective of coat color, the bats might notice light-colored animals at a

TABLE 6
Relationship between Lunar Conditions and Color of Bitten Animals

Coat color	Number of bitten animals	
	Full moon	New moon
White	20	31
Other	14	18

greater distance (using vision for long-distance prey location) and then select individual prey irrespective of coat color when they are close to a herd. According to Suthers (1970), the ability of Microchiroptera to detect small differences in brightness has not been quantitatively measured; tests indicate at least crude brightness discrimination. Evidence of visual form discrimination has been found in two echolocating species (Suthers, Chase, and Bradford 1969). *Desmodus* has an acute ability, relative to other bats, to resolve moving, equal-width black-white stripes (Suthers, Chase, and Bradford 1969). Chase (1972) found that those bat species feeding on large objects have relatively well-developed vision and faint sonar cries. *Desmodus*, with a visual acuity angle of 0.7 degrees, should be able to detect a single cow at 130 m.

Ditmars and Greenhall (1935), Greenhall (1972*a*), and Schmidt, Greenhall, and Lopez-Forment (1970) reported that vampires often returned to the same cow from which they fed previously, and Dalquest (1955) found that vampire wounds characteristically had blood drippings. To see if the odor of blood on a cow influenced prey location, I secured blood from a local slaughter house, defibrinated it, and rubbed it on the necks of twenty-five white Brahma cows. I rubbed water on the necks of another twenty-five white Brahmas, the control animals. All fifty cows were then released into the same pasture, and the brand number of each bitten animal was recorded over the next two days. Four replicates of this experiment were conducted during the dry season, and the number of bitten control animals was compared with the number of bitten blood-marked animals. The results of this experiment are given in table 7. There was no significant difference between the numbers of bites on experimental and control animals ($\chi_1^2 = 0.0$, $P > 0.9$). Vampires do not select individual prey out of a herd because they have the odor of blood on their coats.

Although this experiment did eliminate one factor

TABLE 7
Total Number of Bitten Blood-Marked and Control Animals on Eight Nights

	Trial								
	1	2	3	4	5	6	7	8	Total
Blood-marked	4	2	3	1	2	1	2	0	15
Controls	3	0	1	0	2	3	2	4	15

that could account for the bats' return to the same hosts (the odor of blood), I cannot say whether the odor of blood assisted herd location by the vampire. Given the estimated relative volume of brain associated with olfaction (Mann 1960) and the presence of a vomeronasal (Jacobson's) organ, I suspect that *Desmodus* relies heavily on smell for various aspects of its behavior, including prey location and selection. Wimsatt and Guerriere (1962) reported that, after vampires fed, urine excretion was copious and highly concentrated, and Greenhall, Schmidt, and Lopez-Forment (1971) observed vampires urinating on their prey. Perhaps this is a basis for prey location and selection. Schmidt and Greenhall (1971) and Schmidt (1973) found that *Desmodus* was capable of olfactory orientation in the laboratory and also suggested that it may use smell to find its prey.

HUNTING RANGE AND FORAGING TIME

For the study of hunting range and foraging time, I defined hunting range as the area utilized by the local vampire population as a whole for securing supplies of blood. No estimates of vampire hunting range have been published to date; no differences between the ranges of male and female vampires have been reported. Before com-

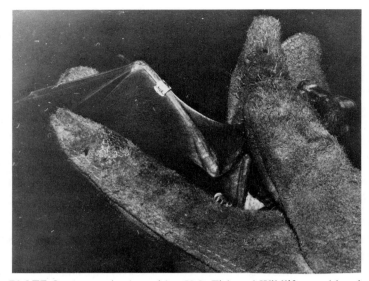

PLATE 8. A vampire bat with a U.S. Fish and Wildlife metal band on the forearm. Using such numbered bands, the author was able to follow the movements and reproductive condition of individual vampires over the fifteen-month study. Band numbers and pertinent information are registered in Washington, D.C., with the U.S. Fish and Wildlife Service.

paring the recapture distances of males with those of females, I compared their rates of recapture. Table 8 lists the total number of male and female vampires banded and recaptured (recaptured only once and recaptured twice) during the fifteen-month La Pacifica study. Chi-square tests revealed no significant difference in recapture rate between males and females ($\chi_2^2 = 0.36$, $P > 0.5$, using only one-time recapture data; $\chi_2^2 = 2.47$, $P > 0.2$, using multiple as well as one-time recapture data); rather, these data indicate relative stability of males and females in the La Pacifica population. If predation on the bats is a significant population-culling factor, it must be about equal on male and female vampires. Overall recapture rate was 28.3 percent (79 one-time recaptures among 279 banded bats).

Whether one sex has a longer life span is difficult to say. Trapido (1946) reported that seven of the eight oldest captive vampires (those living more than eight years) were males. Linhart (1973) calculated average ages of wild caught vampires on the basis of incremental growth-line deposition on teeth. In a sample of 87 Mexican vampires he found a mean age (both sexes) of 2.6 years; for 56 females 3.0 years; and for 25 males, 1.5 years. He concluded that differential survival was suggested. From recapture-rate data, R. J. Burns (personal communication) concluded that there was no sex-related mortality in his Mexican study site.

Now that we know that male and female vampires are recaptured at about the same rate, we can compare the sexes with respect to recapture distances, distances between two capture points for the same individual. These distances were estimated to the nearest 10 m and included zero measurements. Forty-nine recapture distances for each sex were used in the calculations; recaptures within the same month were excluded. A chi-square test for independence on the data in table 9 revealed no significant difference in recapture distance between male and female

TABLE 8
Recapture Rates over Fourteen Months of Male and Female Vampires

Class	Males[a]	Females[a]
Total number banded	146	133
Total number recaptured once	45	34
Total number recaptured twice	12	5

[a]Recaptures within the same month are excluded.

TABLE 9

The Number of Male and Female Vampires in Each Recapture-Distance Category

Sex	Recapture-distance category (meters)			
	0^a	1–250	251–500	500 or more
Males	27	9	3	10
Females	29	5	7	8

[a]Recaptured within 10 m of the first capture point.

vampires ($X_3^2 = 3.04$, $P > 0.5$). The mean recapture distance for male vampires was 366 m ($n = 49$, $SD = 804$, $SE = 116$) and for female vampires, 418 m ($n = 49$, $SD = 874$, $SE = 126$). The mean for both sexes using my data was 392 m ($n = 98$).

Since virtually all of the known La Pacifica vampire roosts were along a river (see figure 3), I calculated the bite rate for each of the pastures to discover whether distance from the nearest river point was correlated with bite rate. Bite rate was calculated as total bites made in each pasture per animal-days for each pasture per year times 1,000. (Since the bite incidence was so low, I used the factor of 1,000 to yield more manageable numbers.) In table 10, I present a distance to nearest river point in standard units and a bite rate for each La Pacifica pasture. A Spearman-Rank correlation test on these data gives a significant negative correlation between distance from pasture to nearest river point and bite rate ($r_d = -0.27$, $N = 46$, $P < 0.05$). Data points were fit to a least-squares equation, and the predicted distance at which bites dropped to zero was interpreted as half the width of the hunting range. Figure 8 is a plot of these points and the least-squares regression line; the distance where bite rate dropped to zero is shown to be 57.1 standard units, or 1.9 km.

From a combination of recapture-rate data and the relationship between distance from a river and bite rate, I believe vampires on La Pacifica forage within a relatively small range consisting of a band 4 km wide, or about 2 km on either side of and along the rivers. This estimate of foraging range is consistent with other indications in the recent literature. Although they did not fix a value to their

TABLE 10

Distance to Nearest River Point and Bite Rate over Fourteen Months for Each Pasture

Pasture number	Distance to river[a]	Bite rate × 1,000	Pasture number	Distance to river[a]	Bite rate × 1,000
2	3	14.07	24	10	1.68
3	10	0.00	25	11	0.00
4	19	1.52	26	8	2.21
5	8	4.04	27	25	2.47
6	32	1.07	28	19	0.35
7	18	6.32	29	18	0.80
8	4	10.17	30	7	0.93
9	30	1.28	31	3	0.00
10	19	1.97	32	3	2.56
11	41	0.00	33	6	3.50
12	41	1.96	34	20	2.31
13	40	1.64	35	35	2.24
14	40	3.95	36	36	0.99
15	47	0.33	38	50	2.73
16	46	0.55	39	45	0.00
17	30	1.67	40	56	0.00
18	24	1.49	41	47	1.35
19	16	0.00	42–46	47	2.10
20	12	1.05	Motel	5	11.83
21	5	5.07	Canabos	20	1.66
22	9	2.96	Cepito	13	0.00
23a	7	1.85	R. Blanca	21	0.00
23b	6	2.71			

[a]Distances are in standard units; one unit = 33.5 m.

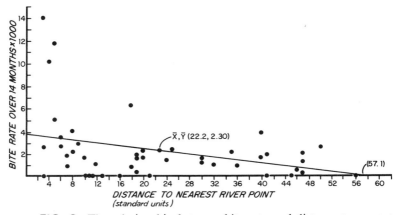

FIG. 8. The relationship between bite rate and distance to nearest river point. The least squares regression line is also shown. (1 standard unit = 33.5 m)

estimate, Fleming, Hooper, and Wilson (1972) did conclude that *Desmodus* had a relatively small range compared with other La Pacifica species and reported no movements between the Tenorio-Tenorito site and any of the Corobici sites, 3–4 km away. In 15 months of banding 279 vampires, I only recaptured two individuals on a river other than the one from which I originally banded them. Thus, Fleming (personal communication) and I believe that there are two subpopulations inhabiting La Pacifica, one roosting along the Tenorio-Tenorito River and foraging in that area and the other roosting and foraging along the Rio Corobici. G. C. Mitchell (personal communication) has also concluded, on the basis of radio telemetry data, that in Mexico *Desmodus* has a small foraging range. Young (1971) saw marked vampires feeding on cattle up to 1.5 miles from their roosts and found that, given sufficient time, they could locate and return to their roost from an experimental release point 10 miles away. Schmidt, Greenhall, and Lopez-Forment (1971) recaptured only 4 of 170 banded bats in different roosts up to 5 km away. R. J. Burns and R. F. Crespo (personal communication) found that interchange between colonies only 2 km apart was significantly greater than interchange between colonies 3 to 5 km away. The foraging range of vampires probably varies from individual to individual and from habitat to habitat.

Without radio-tracking the foraging flights of vampires, it is difficult to estimate accurately the foraging range or the foraging period for an individual bat, which would include travel time from and to a roost, search and detection time, and feeding time. One can cite recaptures on the same night as an indication of the length of the foraging period. Time intervals between successive recaptures of individuals known to have fed and weight changes or stomach code changes between successive captures are listed in table 11. There was no significant difference between the mean time interval for La Pacifica and Palo Verde bats (t-test $= 0.25$, $P > 0.6$); thus the results can be

TABLE 11

Time Interval and Weight Change (or Stomach Code Change) between Successive Captures of the Same Individual on a Night

Interval (hours)	Study site	Weight change (g) or stomach code change[a]
1.3	La Pacifica	2–3
2.0	La Pacifica	+2g
8.0	La Pacifica	+8g
2.0	La Pacifica	+7g
3.2	La Pacifica	+9g
1.8	Palo Verde	1–3
1.8	Palo Verde	+3g
2.1	Palo Verde	+5g
6.0	Palo Verde	+6g

Note: Data are presented only for bats known to have fed.
[a]Code 1 = empty; code 2 = undetermined; and code 3 = full.

combined to give a mean foraging period of 3.1 hours ($SD = 2.2$, $SE = 0.77$).

The real foraging time should be somewhat greater than 3.1 hours, since nets were not necessarily located at the roost, and additional time would be required to move between the netting site and the roost. Wimsatt (1969) found a mean foraging time of about 2.0 hours actually netting the bats at a Mexican cave roost. Greenhall (1972a) stated that selecting a suitable host, biting, and feeding required about 2.0 hours. Based on exit schedules of individuals from roosts and the time of appearances of vampires at cattle, Young (1971) calculated that 1.0 hour was required for finding the prey during the wet season and 2.4 hours during the dry season. I could not analyze my foraging time data by season because of small sample size. If, however, one averages Young's search and detection times (a minimum of half of the total flight time) and multiplies this by two (for the return flight), the total average flight time must be around 3.4 hours, which is very close to the 3.1 hours estimated from my data.

If the vampire is going to be a selective forager, one must assume that it has a high enough prey encounter rate during this three-hour flight to pass over undesirable hosts. Laing (1938) gives an equation for the expected number of food items encountered per unit time, N_t, assuming stationary prey, random search by the predator, and no depletion of prey by the predator:

$$N_t = 2 \; V \left(\sigma_1 + \sigma_2\right) D,$$

where V is the speed of the predator, σ_1 is the radius of the perceptual field of the predator, σ_2 is the radius of the prey, and D is the density of prey.

The assumptions of this equation can relatively safely be applied to vampire foraging behavior. Since vampires locate and feed on sleeping prey, the prey animals are stationary, and there is no chase effort involved. Vampires do not appear to learn the location of herds; they most probably search the pastures for prey in an unsystematic fashion. Lastly, unless the vampire is transmitting a lethal virus, it does not deplete the prey population when it feeds.

To calculate N_t for La Pacifica vampires, I used the following estimates: (1) Let $V = 18$ km per hour. This is about the flight speed of *Eptesicus fuscus* (Layne and Benton 1954), a bat of similar size. (2) Since herds sleep in tight clusters, consider a herd as one prey and point in space. The radius of an average-sized sleeping herd on the ranch is about 10 m, or σ_2. (3) Assume that the radius of the perceptual field of the vampire is 100 m, a conservative estimate, given that the bat need only turn its head or body axis through a maximum of 180° to survey an entire pasture. This would only require seconds, given the visual acuity of the eye (Chase, personal communication). (4) There are an average of fifteen sleeping groups per night using approximately 1,300 ha of grazing land on La Pacifica. Therefore, $D = 15$ per 13,000,000 m^2. According to Laing's equation, a vampire should encounter 4.6 herds per hour of search time. If I assume that 2 of the mean 3 hours between recapture of full bats are used for searching and feeding, a La Pacifica vampire should encounter 9.2 herds per foraging flight, or more than enough to allow selectivity. (Prey selectivity will be demonstrated in chapter 5.)

SUCCESS OF THE HUNT

Mobility (Goss-Custard 1970; Pritchard 1965), prey density (Goss-Custard 1970), and predator experience or conspicuousness (Dixon 1959) have been suggested as influences on the success of various predators. Although Schoener (1971) concluded that the most important factor determining the success of many predators was the relative size of the feeder and the food, this relationship could not exist for *Desmodus*, given the peculiarity of its feeding

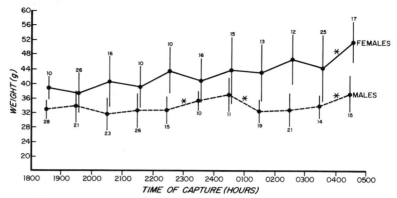

FIG. 9. Weight of vampires at time of capture for males and females. Dark circles designate the mean weight of bats captured during each time interval. Vertical lines through each mean show the mean ± one standard deviation. Numbers above and below vertical lines indicate sample size (*n*).

habits. Likewise, it is difficult to formulate a success/ attempt ratio without assuming that all flying vampires have as their primary goal the acquisition of blood meals. This is probably a safe assumption, given the high energetic costs of flight and maintenance (McNab 1973) and the tight water budget the bats must manage (McFarland and Wimsatt 1969).

Most authors have agreed that considering the amount of time required to feed and return to the roost, *Desmodus* is very efficient at finding a blood meal (Dalquest 1955; Wimsatt 1969; Young 1971; Greenhall 1972*a*). However, little literature on Chiroptera even mentions sex-related differences in foraging behavior.

I followed two procedures to assess the vampires' success at securing their blood meals on La Pacifica and to discover whether one sex was more efficient at foraging than the other. First, I weighed each bat after removing it from the capture net. Figure 9 illustrates the weight at capture for male and female vampires throughout the night.

All data from bats captured under all moon conditions and both seasons were pooled. Females are obviously heavier than males. Significant differences between hourly intervals are marked with an asterisk (*t*-tests for sample means, $P < 0.05$), which may indicate differences in the feeding activities of male and female vampires. The curve for males has three significant changes in weight, while the curve for females has only one. Looking again at the nocturnal activity patterns on new-moon nights (see figure 4), I found no major difference between males and females. However, those activity patterns indicate only when the bats were out of the roost and not *what* they were doing.

To facilitate the collection of data in the field, I used a coding system with the following definitions: Code 1 described the bats whose stomachs were empty; Code 2, bats whose feeding activities could not be determined; and Code 3, bats which had obviously fed. The mean weight, standard deviation, and standard error for fifteen vampires in each stomach category are presented in table 12. Sample means differ significantly from category to category (*t*-tests for sample means, $P < 0.05$); thus I was able to use the coding system and tallies of number of vampires caught in each stomach category to study hunting success. I defined this as the proportion of vam-

TABLE 12

Mean, Standard Deviation (*SD*), and Standard Error (*SE*) of Bat Weights for Each Stomach Condition

Stomach condition	Mean weight	*SD*	*SE*
Empty	32.9	4.4	1.2
Undetermined	36.3	3.3	0.9
Full	47.9	6.8	1.8

Note: Weights are in grams for fifteen bats randomly chosen in each category.

pires netted by dawn with full stomachs, as opposed to empty stomachs.

The number of males caught for each stomach category was compared with the number of females caught for each category, during four time intervals (1800–2100 hours; 2100–2400 hours; 2400–0300 hours; and 0300–0500 hours). For these comparisons, I pooled data on all vampires, regardless of reproductive condition or when they were caught over the year. Several interesting trends appear when the data are grouped by sex, time of capture and, later, reproductive condition.

Chi-square tests for independence on the data in table 13 revealed that for each time interval there was a significant difference in feeding activities between males and females (2×3 chi-square tests: 1800 to 2100 hours, $\chi_2^2 = 14.02$, $P < 0.001$; 2100 to 2400 hours, $\chi_2^2 = 13.34$,

TABLE 13
Total Number of Male and Female Vampires Caught in Each Stomach Category, Grouped by Time of Capture

	Stomach category		
Time interval	Empty	Undetermined	Full
1800 to 2100 hours			
Males	76	7	3
Females	37	6	13
2100 to 2400 hours			
Males	36	3	13
Females	11	5	21
2400 to 0300 hours			
Males	37	9	11
Females	11	5	23
0300 to 0500 hours			
Males	11	5	15
Females	10[a]	1	35

[a]Nine reproductively inactive females.

TABLE 14
Total Number of Females Caught in Each Stomach Category,
Grouped by Reproductive Condition

| | Stomach category | | |
Reproductive condition	Empty	Undetermined	Full
Inactive	51	14	41
Pregnant and/or lactating	18	3	51

$P < 0.10$; 2400 to 0300 hours, $X_2^2 = 16.74$, $P < 0.001$; 0300 to 0500 hours, $X_2^2 = 8.12$, $P < 0.01$). Looking more closely at the tallies within each time interval, I noted the following trends: (a) between 1800 and 2100 hours, for both males and females, I caught more empty bats than full ones; (b) already in the 2100 to 2400 hours interval, more females were caught with full stomachs than with empty ones; and (c) it was not until 0300 to 0500 hours (dawn) that over 50 percent of the males caught had full stomachs.

I noticed that of the ten empty females in the last time interval, nine were reproductively inactive, that is, not pregnant or lactating. This prompted the examination and comparison of stomach conditions between reproductively inactive and active females. A chi-square test for independence was performed on the tallies in table 14 ($X_2^2 = 18.24$, $P < 0.001$), showing a significant difference between females in the two reproductive states. Over the entire night, more pregnant or lactating females were caught with full stomachs, and more reproductively inactive females were caught with empty stomachs.

In a laboratory study on feeding capacities of vampires, Wimsatt and Guerriere (1962) found that consumption averages did not reflect individual differences based on sex or size. Later, Wimsatt (1969) studied foraging efficiency in the natural setting and concluded that differences in foraging pattern between the sexes were not apparent. Granted that the roost-emergence pattern presented in his table 1 is virtually identical for males and females, I note that during his second time interval (2100 to 2259 hours) more than half of all returning females were caught, and less than half of all returning males were netted. I doubt the statistical significance of this difference, yet I suspect that a trend similar to mine might have shown up had he netted more than five nights.

That pregnant and/or lactating females have a different feeding pattern from inactive females and that all females, regardless of reproductive state, tend to feed earlier than males are extremely interesting findings. These indicate that pregnant and/or lactating females must have a high priority assigned to feeding, and they support McNab's (1973) speculation based on energy equations that near-term females may be forced to forage differently from other females. One might also expect a different feeding pattern for males and females, given the weight difference between the two sexes. Schoener (1971) predicted that an increase in energy required should have the same effect on selectivity as a decrease in food density. Indeed, it will be shown in chapter 5 that prey preferences decrease during the wet season, and in chapter 7 it will be shown that a vampire birth peak occurs simultaneously. Still, results of immunological tests on stomach contents indicate males, inactive females, and pregnant and/or lactating females feed exclusively on domestic stock (see table 15).

CHAPTER 5

PREY SELECTION BY VAMPIRE BATS

For convenience, I have divided this chapter into five parts: (1) I will first report the results of this and other investigations on the frequency of domestic versus wild prey utilization by vampires, and I will also propose two possible, though not necessarily realized courses of events leading to vampire hematophagy. (2) I will then give a more detailed description of the prey selection models used as a basis for my research. (3) With this basic information, we will determine *whether* the assumption is correct that vampires today have a more easily accessible, more abundant source of blood from domestic animals than the native wildlife afforded them. We will accomplish this by analyzing the prey preferences among those domestic animals. (4) I will then report the experiments conducted to determine possible bases for vampire preferences. (5) Lastly, I will discuss the central problem of selectivity versus availability in interpreting these results and the model predictions in general.

DOMESTIC VERSUS WILD PREY UTILIZATION

Goodwin and Greenhall (1961) and Villa-R. (1966) reported on the feeding habits of the three vampire species. *Diphylla ecaudata* and *Diaemus youngi* were found to feed primarily on avian blood, while *Desmodus* utilized the blood of both mammals and birds. Later, Greenhall (1970*a*) used the same techniques as I did for the present study (precipitin and haemaglutination tests) and found that *Diaemus* in Trinidad fed on both mammals and birds, while *Desmodus* in Trinidad and Mexico fed on domestic animals, bovines preferably. He also reported that a few vampires had fed on unidentified wild hosts (sample size, 3,500). One blood meal from a wild squirrel has now been confirmed (Greenhall 1972*b*). Schmidt, Greenhall, and Lopez-Forment (1970) found that 80 percent of the *Desmodus* in their Mexican study area ($n = 98$) had fed on cattle, while the remainder had utilized horses, pigs, or chickens as prey.

Desmodus at La Pacifica and at Palo Verde apparently feed exclusively on the blood of domestic animals, and principally that of cattle and horses. The results of the precipitin and haemaglutination tests on the blood meals of 58 vampires, representing up to 174 feedings, are presented in table 15. None of the samples contained blood from wild prey, nor from domestic animals other than the bovine or horse, regardless of season of capture, the bat's sex or reproductive condition, or lunar phase during capture. Occasionally, a goat or pig was also bitten on La Pacifica, but this occurred so rarely that the data are not quantifiable. I made spot checks on dead chickens from the laying houses (non-vampire-proof cages) and never found vampire wounds. I regularly saw deer crossing pastures near the forest tracts in the early morning; however, I never spotted

PLATE 9. Male vampire bats sacrificed in order to secure blood-meal samples for the precipitin and haemaglutination tests. On the filter paper above the bats are samples of blood taken from a bat's stomach, intestine, and colon; these samples were analyzed in England to determine what the bat had fed on. Note the spherical shape of the male which has fed on the left.

TABLE 15

Results of Precipitin and Haemaglutination Tests
on Vampire Blood-Meal Samples

Conditions of capture	Number of bats sacrificed	Number of feedings[a]	Percentage domestic stock
Wet season	9	27	100
Dry season	5	15	100
New moon	6	18	100
Full moon	3	9	100
Male bats	5	15	100
Inactive females	6	18	100
Pregnant/lactating females	4	12	100
Palo Verde cave bats	20	60	100

[a]Three feedings from each sacrificed bat. Greenhall (1970*a*) determined that by removing blood from the stomach, intestine, and colon, the feeding activities of the three nights prior to capture could be assessed.

any indication of a vampire wound, even using binoculars. K. Glander, who was studying Howling monkey behavior and ecology on the ranch, never observed vampire bites on these wild animals.

Certainly I would expect a difference in the proportion of wild versus domestic prey utilization for different environments, depending on the relative abundance of each prey type. However, it remains that within each study area where vampire feeding behavior has been examined (several locations in Mexico, Trinidad, and Brazil, for instance) and even within my Costa Rican study site, where at least some wild prey species were available, *Desmodus* feeds almost exclusively on domestic stock.

This implies that within the last four hundred years vampires have altered their feeding habits significantly. In Pre-Columbian times, there were relatively few large herding mammals, and vampires may have had problems finding adequate food (McNab 1969). *Homo sapiens* may have been an intermediate host between wild and domestic animals. Benzoni (1967–[1565]) and De Oviedo y Valdes (1950–[1526]) reported that humans were frequently attacked during the Conquest; even today humans are sometimes fed upon (Greenhall 1970*a*, as well as my own personal observation). Small villages of potential human hosts were located along Guanacaste rivers over two thousand years ago (D. Stone, Muséo Nacional de Costa Rica, personal communication). When cattle and horses were brought to the New World, they were corralled close to the villages. If vampire populations were already concentrated in village areas, the newly imported domestic animals would make readily accessible targets. Selection could have favored vampire populations utilizing these concentrated prey centers, which may explain why Fleming, Hooper, and Wilson (1972) found *Desmodus* to be relatively uncommon in undisturbed Panamanian forests. De Azara (1935) reported that by A.D. 1700 there were already forty-eight million head of *wild* cattle inhabiting the South American campos and pampas from 26°S to 41°S latitude! Four hundred years is a short period in an evolutionary context; learning may have been a prime factor in the switch to domestic prey. (I will present evidence later that young vampires learn how and what to feed on [see chapter 7].)

Another possible and more plausible pathway by which vampires evolved their curious feeding habits is shown in figure 10. Bats evolved from the early insectivores (Romer 1959), and even today many species feed on insects. Perhaps a progenitor of today's vampire specialized on the ecto-parasites of larger wild mammals. Buettiker (1959) and Baenziger (1968) have shown that blood-sucking moths feed on domestic and wild hosts, and an eye-frequenting moth has been reported disturbing horses in Argentina (Shannon 1928). Buettiker (personal communication) has

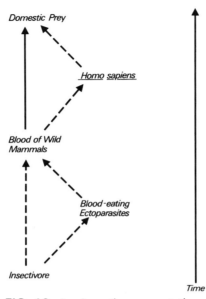

FIG. 10. A schematic representation of possible evolutionary pathways leading to vampire hematophagy and current prey preferences.

observed bats feeding on parasitic moths of cattle and wild hosts in Africa, and he will soon investigate this phenomenon in Brazil. Arata, Vaughan, and Thomas (1967) found that 17 percent of the 23 *Desmodus* they captured in Columbia contained insects in their stomachs. Perhaps vampires first fed on the blood-feeding ectoparasites of wild hosts and later directly on the mammalian blood. The transition to domestic stock could still have involved *Homo sapiens* as an intermediate host.

PREY-SELECTION MODELS

It has often been assumed that vampires today have a more abundant source of blood from domestic animals than the native wildlife afforded them (Dalquest 1955; Greenhall 1968, 1972*b*; Villa-R. 1968). The two theoretical models discussed below predict that the higher the absolute abundance of food, the narrower the range of items in the diet. In other words, the more food available, the greater selectivity a predator can exhibit. Therefore, if *Desmodus* can be shown to be highly selective within the various domestic stock types, the predictions of both models will be supported, and we can state factually that the bats have an abundant source of food in domestic animals.

The MacArthur and Pianka Model. MacArthur and Pianka (1966) and MacArthur (1972) concern themselves with the range of dietary items natural selection should favor if the animal feeds optimally. Their basic approach to optimal utilization of time and energy is quite straightforward: any foraging activity that results in a *net* energy gain should be increased. Different food types have a lower cost of securing and eating than other food types, relative to the food energy gained. They assume that these food items can be ranked from most profitable to least profitable and that the feeding animal can somehow assess the potential profit. This is probably not an unreasonable assumption, given that a feeder could learn even by trial and error which food types are more difficult to find or take longer to capture or eat. They also assume that during the search for food the prey species are discovered in the proportion in which they occur. The animal's diet should be increased to include those items offering a net energy gain; once the ranked food item for which the cost is greater than the gain is reached, no further diet enlargement should be made.

MacArthur and Pianka (1966) express these relationships mathematically as follows: assume that a predator already includes N kinds of prey in its diet. They subdivide its time per item eaten, T_N, into a search time, T_N^S, and a pursuit (plus capture and eating) time, T_N^P. If the diet is enlarged to include $N + 1$ items, one can find the change in total time, ΔT_N, which accompanies enlarging the diet.

MacArthur and Pianka rank the items from highest harvest per unit time to lowest and proceed through the list of items until $\triangle\ T_N$ first becomes negative. Change in search time, $\triangle\ T^S$, is always negative because the more acceptable items a predator has, the less the search time per unit of food. However pursuit time, T^P, would increase as harder-to-catch items are added to the diet. If one plots the reduction in search time, $\triangle\ S$, and the increase in pursuit time, $\triangle\ P$, for a hypothetically ranked list of different food items, the point of intersection shows the optimal number of food types in the diet. In such a graph, higher prey density decreases the reduction in search time, $\triangle\ S$, and results in an intersection point along the dietary axis closer to the origin (higher selectivity).

The Emlen Model. Emlen (1966) arrives at the same conclusion from a probabilistic approach. He assumes that natural selection will favor the development of feeding preferences that will, within certain limits, maximize the net caloric intake per individual per unit time. Given two kinds of items, *i* and *j*, he asks what fraction of the time the predator should stop and eat rather than skip the item (*i* or *j*) before moving on. He also assumes that this fraction is the probability that the first alternative offers a greater net energy per feeding time than the second and that the predator knows in advance the net energy and feeding time for the next item (Schoener 1971). The second item encountered is always assumed to be consumed. From Emlen's equations it is possible to predict how changes in abundance, both relative and absolute, affect the proportion of *i* and *j* in the diet. An increase in prey density means a decrease in the distance between items; the probability that the predator remains to feed on an item approaches zero, and the predator becomes more selective.

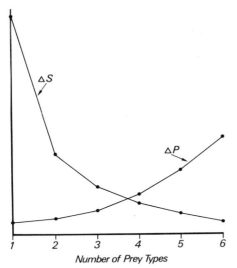

FIG. 11. The relationship between the reduction in search time, ΔS, and the increase in pursuit time, ΔP, accompanying changes in the number of prey types included in the diet. [After MacArthur and Pianka (1966)].

PREY SELECTION AMONG DOMESTIC STOCK

Although a goat or a pig was occasionally attacked by a vampire, the frequency of such feedings was so low that I excluded these domestic animals from my examination. Likewise, I excluded the large chicken population after finding no evidence of vampire attacks on dead birds and no vampire blood meals positive for avian blood by the precipitin test. La Pacifica vampires primarily attacked cattle and horses. Horses were the most preferred domestic species on the ranch; yet a total population of from thirty to thirty-five horses was hardly enough to serve as the focus for my research over fifteen months. Thus I chose the large cattle population in order to assess vampire prey preferences and the potential bases for those preferences.

I determined vampire preferences among the various breed, age, and sex classes of bovines in the following way. During each of six two-month sampling periods, I totaled the number of bites on the Swiss and Brahma breeds and compared these with expected values, corrected for the number of each breed available as potential prey. I also calculated a corrected ratio for breed preferences by dividing the number of bites per animal for Swiss by the number of bites per animal for Brahma during each of the sampling periods. Animals of all ages and both sexes were included in the breed-preference test. Since each herd on the ranch contained both breeds, the position of the herd relative to vampire roosts (or rivers) would not affect breed preference, and bite data could be pooled over the entire ranch.

The sample periods are presented in chronological order in table 16; wet season occurred from mid-May to mid-November 1972. For each two-month sample period, vampires significantly preferred the Swiss over the Brahma animals (chi-square tests for goodness of fit, $P < 0.05$). However, looking at the corrected ratio of bites in this table,

TABLE 16

Vampire Preference for the Brown Swiss Breed on La Pacifica

Season of sample period[a]	Direction of preference	Corrected bite ratio[b]	n (bites)
Dry	Swiss over Brahma	9.08	99
Wet	Swiss over Brahma	4.26	96
Wet	Swiss over Brahma	4.11	142
Wet	Swiss over Brahma	2.75	240
Dry	Swiss over Brahma	6.24	197
Dry	Swiss over Brahma	10.81	149

[a]Sample periods are presented in chronological order; the wet season was from mid-May to mid-November, 1972.
[b]Ratio of bites on Swiss : Brahma stock calculated by the method described above.

the degree of preference for Swiss over Brahma decreased during the wet season, a point to which I will return later.

To determine whether vampires preferred calves (of either sex) over their cows, during both the wet and dry seasons I sampled and compared the bite distribution on calves and cows where the number of each available prey type was known for a given pasture. On La Pacifica, there were herds containing both calves and their cows and herds of cows only. Since I had previously shown a relationship between bite rate and distance to the nearest river, it was necessary to compare the bite distribution between calves and cows in the same pasture. I tested the existence of a sex preference among calves in the same manner, with the number of bitten male and female calves and the total number of each sex available to choose from known. Corrected ratios of these preferences were calculated. Chi-square tests on the data in table 17 show (1) a clear preference for calves over their cows in the same pasture ($P < 0.05$); (2) no sex preference among the calves themselves ($P > 0.05$); and (3) no seasonal trends with respect to calf preferences.

TABLE 17

Vampire Preferences among Calves (Weaned at Eight Months) on La Pacifica

Conditions of observation[a]	Corrected bite ratio[b]	n (bites)
Wet season—calves : cows	3.5	36
Dry season—calves : cows	3.2	42
Wet season—female : male calves	1.09	22
Dry season—female : male calves	1.11	31

[a]Samples were taken during both seasons when the number of calves, the number of cows, the number of male calves, and the number of female calves were known for each pasture.

[b]Calf : cow and female calf : male calf ratios were calculated by the same method as for breed preference.

PLATE 10. Heifers of the bovine breeds on La Pacifica. The two breeds are maintained together in each of the ranch's herds. *Above*, a typical Brown Swiss heifer, with nearly horizontal ears; *at right*, a Brahma heifer, with characteristic drooping ears. The Swiss breed animals were preferred by vampires at all times of the year.

I determined vampire sex preference among adult bovines in the same manner as that used to determine breed preference, excluding bites on calves, of course. Except for the breeding herds, adult males and females are maintained in separate herds and rotated from pasture to pasture. No attempt was made to keep herds of either sex in pastures a certain distance from any river system. Therefore, I pooled the bite data within each sex, regardless of where the herd was located. The two-month sample periods are presented in chronological order in table 18; chi-square tests revealed a significant preference for adult female over adult non-

castrated male bovines throughout the year ($P < 0.05$, for each sample period). Again, during the wet season, there was a pronounced decrease in the degree of preference for females (lower ratios), which will be discussed in the next section.

POSSIBLE BASES FOR VAMPIRE
PREY PREFERENCES

Breed Preference. As the name implies, Brown Swiss have a tendency to be dark animals. To determine whether vampires prefer dark or light-colored cattle, I conducted

TABLE 18
Vampire Preference for Adult Female Bovines on La Pacifica

Season of sample period[a]	Direction of preference	Corrected bite ratio[b]	*n* (bites)
Dry	Female over male	15.6	64
Wet	Female over male	4.1	66
Wet	Female over male	4.4	94
Wet	Female over male	6.8	176
Dry	Female over male	8.9	138
Dry	Female over male	12.4	94

[a]Sample periods are presented in chronological order; the wet season was from mid-May to mid-November, 1972.
[b]Female : male ratios of bites were calculated by the same method as that for breed preference.

two experiments. First, I removed all of the Swiss breed from a herd, leaving only Brahmas of various colors in the pasture. The numbers of bites on dark and light animals were compared, again accounting for the difference in numbers of dark and light animals available. Second, I was fortunate to have three white Swiss heifers on the ranch. I placed these three with three white Brahma heifers in a separate, small pasture and compared the number of bites on each breed.

The results of the two color-preference tests are given in table 19. In the pure Brahma herd of mixed coat colors, there was no significant preference for dark-colored animals (chi-square test, $P > 0.1$). Looking at the bite ratio for the three white Swiss and the three white Brahmas, the vampires still preferred the Swiss over the Brahmas (chi-square test, $P < 0.05$). Thus coat color alone is not enough to explain the preference for Brown Swiss. Although a preference for dark-colored animals has been reported in the popular literature (Agricultúra de las Américas 1972), no data have been published to substantiate this. In my experiments, I controlled both breed and coat color.

Perhaps the degree of exposure to potential vampire attack could influence prey selection and breed preference. Since I had already determined that vampire attacks occurred during the herd's resting period, I conducted two experiments to determine whether sleeping position within a herd influenced prey selection.

During the dry season, I approached a sleeping herd on horseback and recorded the overall herd size, the number of animals on the perimeter, the number of perimeter Swiss and central Swiss, and the number of perimeter and central Brahmas. I called an animal "perimeter" if it was on the edge of the herd cluster and if I could see at least one-half of its body from my position outside of this cluster. In this manner I could see if either breed in particular might be more exposed to potential vampire attacks by being on the perimeter of a dense, sleeping herd. The results appear in table 20. I found that when a herd grouped to bed down and sleep, the Swiss members of the mixed-breed herd were con-

TABLE 19

The Lack of Vampire Preference for Animals of Dark Coat Color

Conditions[a]	Corrected bite ratio[b]	n (bites)
1. Same breed—dark : light coat color[c]	1.5	30
2. White Swiss : white Brahmas	6.0	14

[a]For the first experiment, data were from an all-Brahma herd of mixed coat colors; for the second experiment, data were from Swiss and Brahmas of the same coat color in the same pasture.
[b]Corrected bite ratios were calculated by the same method as for breed preference.
[c]"Dark" coat color includes all coat colors but white.

sistently located on the perimeter of the cluster (chi-square test, $P < 0.01$). This was the first indication that exposure to potential attacks might be important in vampire prey selection. But indications are not sufficient proof, and I decided that a further experiment was in order.

I used my bite records to find six, young Brahma males that were bitten repeatedly, or were preferred, by vampires. These were culled out of their herd, marked for long-distance observation with greasepaint, and released back into the same herd. On five sampling nights, or the duration of the grease mark, I recorded the sleeping positions of these six preferred Brahmas relative to other herd members. Table 20 indicates that Brahma individuals known to be preferred by vampires consistently slept on the perimeter of the cluster (chi-square test, $P < 0.05$).

A further indication that exposure to potential attacks is of critical importance can be found in table 21. In the

TABLE 20

Sleeping Locations Within a Herd

	Herd size	Perimeter			Central		Marked on perimeter	Central marked
		Total	Swiss	Brahma	Swiss	Brahma		
Location of Swiss relative to Brahma[a]	87	32	14	18	3	52	—	—
Location of marked, preferred Brahma relative to other Brahma[b]	136	43	—	—	—	—	5	1

[a]Mean values are based on twelve sampling nights.
[b]Mean values are based on five sampling nights.

TABLE 21

Distance-to-Nearest-Neighbor Relationships Within a Herd and During the Two Seasons (estimated in cow lengths to the nearest whole number including zero)

Season	Mean (in cow lengths)	SD	n^a
Dry season	0.30	0.46	40
Wet season	1.20	0.72	40

Note: The same-sized herd was observed under the same lunar conditions during the two seasons and in the same pasture.

[a]Ten distance estimates made on each of four nights during each season.

previous section, I noted that the preference for Swiss over Brahma decreased during the wet season. If the exposure hypothesis is correct, I would expect to find more Brahmas exposed relative to Swiss during the wet season. I compared the distance to nearest neighbor for the same sized herd in the same pasture during the wet and dry seasons. Distances were estimated in cow lengths to the nearest whole number including zero. Table 21 shows that the distance to the nearest neighbor significantly increases during the wet season, effectively lessening the Swiss's hold on perimeter positions and creating more Brahma target animals (*t*-test for sample means, $t = 6.67$, $P < 0.05$).

Since vampires bite when the herd is bedded down, one might look for differences in the amount of time each breed spends sleeping (or, conversely, grazing) each night. As mentioned earlier, the breeds are maintained in mixed herds. On the basis of thirty-six nights of observation during both seasons, I can state that, without exception, there is no difference between Swiss and Brahma breeds in the amount of time spent grazing at night. The herd behaves as a unit; when the first animal rises to graze or lies down to sleep, the entire herd does so, regardless of breed. Thus, nocturnal

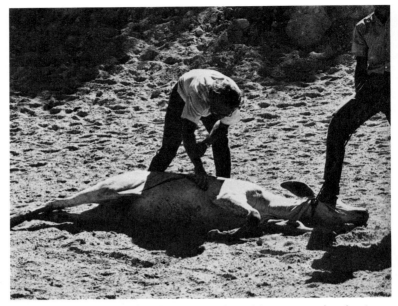

PLATE 11. Marking Brahma males with grease paint for long-distance identification. By marking preferred individual Brahmas, the author was able to determine their sleeping position relative to other herd members and discovered that they regularly sleep in exposed positions.

grazing differences could not account for the vampire's preference for Swiss animals. Lampkin and Quarterman (1962) and Moran (1970) also found no significant differences in nocturnal grazing time between European and Brahma breeds. Rhoad (1938) reported that Zebu (Brahma-type) cattle graze longer and travel greater distances than European cattle in the tropics; however, he failed to separate diurnal from nocturnal grazing time in his calculation. Payne, Laing, and Raivoka (1951) found no change in the amount of nocturnal grazing with season.

Schmidt, Greenhall, and Lopez-Forment (1970) reported that Mexican vampires most preferred the Swiss breed, then Charolais, Santa Gertrudis, and lastly Zebu breeds. They speculated that breed preferences may be based on different breed behaviors or on some skin or blood substance attracting or repelling the bats. Albert Schaffer, veterinarian, has suggested two explanations for European over Brahma breed preferences (personal communication). First, the Brahmas have an unusual anatomical structure between the skin (hide) and the underlying cutaneous muscle which allows them to contract the cutaneous muscles violently, causing the skin to twitch. This may be very annoying and disruptive to the vampire when feeding. Either European breeds lack this structure or it is poorly developed. Second, the skin secretions of the Brahma are known to repel flies and mosquitoes. Schaffer also suggested that the body odor of the Brahma may be objectionable to the vampire and result in selection of European breeds. Both explanations seem logical; however, the former would not explain why the degree of preference for European (Swiss) breeds decreases during the wet season. Additionally, horses have the ability to twitch their hides; and they were highly preferred prey on La Pacifica. The odor hypothesis might still be applicable if one assumes that rains eliminate objectionable odors from the hide during the wet season.

I found that Swiss animals bed down more often on the perimeter of a herd than Brahma animals; also, individual Brahmas known from previous bite records to be repeatedly attacked by vampires slept on the perimeter of the herd. These results strongly suggest that vampires inflict their wounds on perimeter animals. This seems logical, if one considers the potential danger to a vampire of being crushed while feeding in the middle of a densely packed herd. But why do the Swiss animals consistently rest on the perimeter of a sleeping herd?

Kilgour and Scott (1959) showed that cows of those middle-dominance rank were in the advance of voluntary herd movements; those of top-dominance rank were in the middle of the herd; and those of low-dominance rank were at the rear of herd movements. If I assume that movement order approximately reflects sleeping positions, then low- and middle-ranking individuals are on the perimeter of the herd, and top-ranking individuals are in the center and least exposed to vampire or other predator attack. Schein and Fohrman (1955), using Jersey and Zebu-Jersey cross animals, found no relationship between breed and social rank. If the breed of animal is not responsible for its social rank and therefore not the reason for Swiss occupancy of perimeter positions, what is?

On La Pacifica, the temperature during the dry season is only from three to four degrees higher than during the wet season. However, nocturnal winds are common during the dry season. Johnston (1963) has shown that high wind velocity reduces the body temperature of cattle most effectively when the humidity is low. Perhaps the Swiss breed, with its darker and thicker coat, is on the perimeter to increase evaporative cooling, or the Brahma breed is central to conserve heat during the night. Whittow (1962) found that increased blood flow to the skin was one mechanism of heat loss in cattle. This could mean that blood flow from a Swiss wound is greater than blood flow from a Brahma wound. Since Voisin (1959) has demonstrated that distances to nearest neighbor are greater at high air temperature, and since I have found an increase in distances to nearest neighbor during the wet season (table 21), the cooling effect of the nocturnal, dry-season winds may be significant enough to influence position within a sleeping herd.

Calf-Cow Preference. I compared calves and cows on two parameters related to the degree of exposure to potential vampire attacks. First, I recorded the number of one-hour periods through the night when over 50 percent of the calves were standing or walking and the number of periods when over 50 percent were sleeping. These data were also recorded for cows in the same pasture at the same time. Second, I compared the distance to nearest neighbor when calves were in a pasture with their cows with the distance when calves were absent but replaced by an equal number of cows. Calves were treated as if they were cows in this test. These observations indicate that calves are more exposed to vampire attacks than cows. Table 22 shows that even when the herd is bedded down, the distance to nearest neighbor is greater when calves are present in a herd than when they are not (t-test for sample means, $t = 8.28$, $P < 0.05$). Table 23 indicates that calves spend a significantly greater part of the night bedded down than do their cows in the same pasture ($\chi_1^2 = 10.6$, $P < 0.01$). Hafez, Schein, and Ewbank (1969) previously noted this fact.

Dalquest (1955) and Linhart, Crespo, and Mitchell (1972) also found that vampires in Mexico preferred calves over adult bovines. Thickness of hide might influence this preference; still, the observations that calves spend more time sleeping at night than adults and that they sleep at greater distances from nearest neighbors than adults indicates that degree of exposure to potential attackers may be the important factor in prey selection.

TABLE 22
Distance to Nearest Neighbor for a Herd of Cows Only and an Equal-Sized Herd of Cows and Their Calves (estimated in cow lengths to the nearest whole number including zero)

Animals observed[a]	Mean (in cow lengths)	SD	n[b]
Cows only	0.33	0.47	40
Cows with calves	1.35	0.62	40

[a]The same-sized herd observed under the same lunar conditions.
[b]Ten distance estimates made on each of four nights for each herd type.

TABLE 23

The Amount of Time Spent "Sleeping" during the Night by Calves and by Their Cows[a]

Animals observed	Number of 1-hour periods sleeping	Number of 1-hour periods standing
Calves	42	2
Cows	29	15

Note: Data were taken on four nights from calves and cows in the same pasture.

[a]By over 50% of the calves and by over 50% of the cows in the herd.

Adult Sex Preferences. Given that degree of exposure to potential vampire attacks could be related to both the breed and age (calf versus cow) preferences of vampires, it was necessary to demonstrate that the same underlying mechanism held for the adult sex preference. I made the same observations on herds of each sex on the same nights (males and females are maintained in separate herds on La Pacifica). I compared the number of one-hour periods spent sleeping (that is, by over 50 percent of the herd) and the number of periods spent grazing by a young male herd and a heifer herd. Distance to nearest neighbor, in cow lengths, was also recorded and compared (these data are presented in tables 24 and 25). Differences between males and females in both tests were not significant (time spent bedded down, $\chi_1^2 = 0.04$, $P > 0.2$; distance to nearest neighbor, t-test for sample means, $t = 0.24$, $P > 0.1$). If the female bovines maintained greater distances to nearest neighbors or spent more time sleeping at night, I would expect a preference based on exposure. I found no significant differences between males and females on either measure; thus it appears that exposure may not be the mechanism governing vampire preferences for adult female bovines. I will return to this point shortly.

TABLE 24
The Amount of Time Spent "Sleeping" during the Night
by Male Herds and by Female Herds[a]

Animals observed	Number of 1-hour periods sleeping	Number of 1-hour periods standing
Male herd	41	14
Female herd	39	16

Note: Data are from the same five nights for approximately equal-sized male and female herds in different pastures.
[a]By over 50% of the herd.

I also attempted to determine whether an odor was responsible for the sex preference of vampires. On three occasions I swabbed the vulval-anal odor of females onto twenty-five males and water onto another twenty-five males. The experiment was a failure in that none of the fifty were bitten during any of the trials. However, I conducted the experiment incorrectly in the first place; I did not know the estrous condition of the cows utilized.

Toward the end of my fifteen-month study I decided to see if vampires preferred estrous or non-estrous cows. This

TABLE 25
Distance-to-Nearest-Neighbor Relationships for a Male Herd and
a Female Herd (estimated in cow lengths to the
nearest whole number including zero)

	Mean (in cow lengths)	SD	n^a
Male herd	0.30	0.46	40
Female herd	0.33	0.47	40

[a]Ten distance estimates made on each of the same four nights for both herds.

presented a unique problem, in that even the experienced veterinarian has difficulty pointing out an estrous cow in the field. I decided to let the bulls do it for me; I constructed and strapped harnesses with paint sponges onto the bulls; whenever a bull mated with an estrous cow, she was temporarily marked. Then each morning I checked the herd for vampire bites and noted whether or not the cow came into estrous the previous night. The number of bites on estrous and non-estrous cows was compared. I should note that the bull/cow ratio was purposely set high to ensure that all estrous cows would be serviced and marked. The results, in table 26, demonstrate that cows in peak estrous are significantly preferred by vampires over cows in other phases of the estrous cycle ($\chi_1^2 = 151.6$, $P < 0.001$). Since this was discovered during the last month of field work, I could observe the nocturnal behavior of four estrous cows only on each of two nights. These estrous cows were marked with grease paint, as described earlier. At all times when the herd was bedded down, each of the marked estrous cows was located on the perimeter of the herd. In addition, I found that they frequently alternated between standing and lying down in that perimeter position, while other herd members remained bedded down.

Uncastrated males have much thicker and tougher hides than cows (A. Schaffer, personal communication),

TABLE 26
Vampire Preference for Cows in Peak Estrous

Estrous phase	Number of cows with vampire bites	Number of cows without vampire bites
Cows in heat	15	7
Cows not in heat	6	322

Note: Data are based on seven nights of observation.

possibly making the cows the preferential hosts. But again, this would not explain the decrease in female preference during the wet season. Given the strong preference for cows in peak estrous and the low number of bitten animals on the ranch, I suspect that most of the adult females bitten were in estrous and that the "female" preference was really a preference for estrous cows. Assuming this to be correct, then all three vampire prey preferences—for Swiss over Brahma stock, for calves over cows, and for adult females over adult males—can be linked to the degree of exposure to vampire attacks. Swiss breed animals are more exposed than Brahma stock because they sleep on the perimeter of a resting herd; calves are more exposed than their cows because they sleep for longer periods during the night and with greater distances to nearest neighbors; and adult females are preferred over adult males most probably because estrous cows are highly preferred and are also more exposed, being on the perimeter of the herd.

This last preference is curious, in that estrous cows are hyperactive (Hafez, Schein, and Ewbank 1969; personal observation) and alternate between standing and reclining while other herd members are sleeping. Since vampires inflict their wounds when the herd is bedded down, these animals would be more easily spotted but possibly difficult to feed on. A. van Tienhoven (personal communication) suggests that their hyperactivity may distract their attention from vampire attacks, and the bats might be able to obtain blood relatively undisturbed. He also suggests that there might be selection for estrous cows because they have low reproductive hormone concentrations at peak estrous. At other phases of the cow's cycle the progesterone levels are higher. The estrone level is higher just prior to estrous, and estradiol peaks at estrous (Hansel and Echternkamp 1972). These might have an adverse effect on the vampire's reproductive system if they are absorbed without being broken down. Similarly, high amounts of testosterone in the

PLATE 12. The method used to identify peak-estrous cows in the field. *Above*, a Brahma bull with the paint harness attached to his face; *at right*, an estrous cow marked while mating the night before. Notice the dark paint spots on her dorsal side. Vampire bats significantly preferred estrous cows as prey.

blood of males might have inhibitory effects on the pituitary gland. If the vampires are sensitive to these hormones, a substantial intake with the blood might well have the effect of a birth-control pill. The preference for calves would also eliminate hormone-intake problems. Mead-Briggs and Rudge (1960), Rothschild (1965), and Foster (1969) have already demonstrated how the life cycles of different parasites can be controlled by the hormone cycles of the hosts.

Estrous cycling might influence prey preferences in two additional ways. First, since female preference decreases during the wet season, there might be a difference in the number of estrous periods during the two seasons. Plasse, Warnick, and Koger (1970) found no significant difference in the number of estrous periods per heifer per year between seasons. Also, on La Pacifica there is no

attempt to breed cows on a seasonal basis (see table 1), so
the number of cycling cows remains about the same in both
seasons. Second, Brakel, Rife, and Salisbury (1952) found
that the intensity of estrous is less marked in Brown Swiss
than in other breeds. One measure of intensity was the
restlessness of the cow. Thus estrous Swiss cows are not
only on the perimeter of a sleeping herd; they are probably
less restless than estrous Brahma cows on the perimeter.
Hafez, Schein, and Ewbank (1969) have also shown that the
duration of estrous is longer in European breeds than in
Zebu (Brahma) stock, and both the intensity and duration
of estrous may influence vampire breed-estrous preference.

Here I must add another bit of speculation regarding
estrous-cow selection, which is intended only to stimulate
future research on this topic. In chapter 3, I estimated that

a mean 10.5 bats utilized the same prey animal on any given night, taking available foraging time into account. With each vampire consuming an average 15 ml of blood per feeding (McFarland and Wimsatt 1969), a cow could lose 1.5 L of blood each night. Over several nights, with continued utilization, detrimental effects would certainly appear. Greenhall (1970*b*) has observed these debilitating effects in areas with severe vampire problems. With selection for estrous cows, and with the cows cycling only once every 20–21 days (Asdell 1964), such a system, by spreading the population's blood demands around, could prevent debilitation and loss of prey animals. Thus, the La Pacifica vampire-cattle interaction might be more balanced than in other areas where the vampires return to feed on the same animal night after night (Ditmars and Greenhall 1935). In other areas, the bats may be forced to do this because of a high bat/prey ratio.

PREY SELECTIVITY
OR AVAILABILITY?

MacArthur and Pianka (1966) and Emlen (1966) have predicted that animals will be more selective in their diet when food is common than when it is rare. The behavior of *Desmodus* appears to support this prediction. *Desmodus* not only selects domestic over wild hosts but also prefers particular breeds of stock, particular age and sex classes, and individuals in a particular reproductive state. I have presented evidence that degree of exposure to potential attackers is an important basis for prey selection by the vampire. *Desmodus* most frequently attacks animals which sleep on the perimeter of the herd, sleep with greater distances to their nearest neighbors, and/or sleep for longer periods during the night. Additionally, when the overall degree of exposure is increased (distance to nearest neighbor increases during the wet season), specific preferences, which are at least in part based on high exposure, decrease.

Throughout this chapter, I have repeatedly referred to prey "preferences" of the vampire and "exposure" of the prey. Are these bats really selecting or preferring one prey type over another, or are they merely feeding on the only available animals, those which are exposed? A number of observations point to an actual preference or selection process. Greenhall (1972a) has observed that when a variety of hosts (breeds) is present, the vampire is selective; when only the least selected host (Zebu breed cattle) is available, vampires may attack that type severely. I made a number of visits to Finca Carrizal, a ranch with only Brahma stock but surrounded by others with mixed European-Brahma breeds. Small vampire roosts were located in the center of this property. When my less rigorous bite surveys were conducted during moon phases allowing high foraging time, the ranch's Brahma stock remained untouched. The only vampire bites occurring on this ranch were discovered just prior to a full-moon period. Thus it certainly appears that the vampire is preferring certain prey types (flying over the Brahmas) when sufficient foraging time is available and feeding on less desirable types at other times.

Normally during the course of my studies I was not attacked by vampires. Yet on one evening, when all of the cattle had been moved to the far end of the ranch for a health inspection, I became the "selected" host. Quite often when I was sitting near my capture nets, I would hear and see vampire-sized bats circling around me. The bats have been observed doing the same above and around cattle herds (Greenhall, Schmidt, and Lopez-Forment, 1971; personal observation), and this may be part of the selection process. Often predators will test possible prey by chasing it (Mech 1966) or nudging it for juices (Springer 1960); how the vampire selects a particular individual as a host is still undetermined. Greenhall, Schmidt, and Lopez-Forment (1971) have described the reactions of bovines when vampires land on their bodies; the cows shake their heads, flap their ears, or attempt to brush the bats off with their muzzles or horns. During the night when surveying a herd for bites at close range, I occasionally noticed fresh, vampire-sized wounds on bodies without any characteristic blood drippings. Perhaps the bats test the potential prey in this manner; when the bovine reacts strongly to the biting, the bat flies off to find a more agreeable host.

CHAPTER 6

ASPECTS
OF SOCIAL
BEHAVIOR

Gould, Woolf, and Turner (1973) and Schmidt (1972) have reported on the communication calls between mother and infant vampire bats. Both studies describe a two-part "isolation call" given by the young vampire when separated from its mother. In a separate paper, I have discussed individual variation in these vocalizations and their function in individual recognition by *Myotis* (Turner, Shaughnessy, and Gould 1972). In a review of the literature, Gould (1970) found that in most bat species studied, the mother recognized her own young and repelled other infants that attempted to nurse. Greenhall (1965) noted this in *Desmodus*. Schmidt and Manske (1973) have studied the behavioral development of young *Desmodus* in captivity and found that it takes from nine to ten months to reach adult weight. They saw no signs of aggressive behavior in young vampires, which have an appeasement gesture to suppress aggression in approaching adults which consists of lifting one folded wing and bending the body to the side.

Schmidt and van de Flierdt (1973) have given the most detailed descriptions of aggressive behavior in *Desmodus* trained to feed from a bottle on cue. In an agonistic encounter of low intensity, the approaching adult maintains a lower posture, and its neck hairs are bristled. Both bats push at each other side to side. In more intense interactions, a bat may knock one wing on the ground and/or drum both wings rapidly. In an encounter of the highest intensity, or a real attack, there is a frontal approach to 10–15 cm, a quick lunge toward the opponent, drumming of wings, and loud cries. If the opponent does not withdraw fast enough, it is beaten by the attacker's wings. Biting is rare and is only done by an inferior animal lying on its back. Schmidt and van de Flierdt reported that an "anxious" animal assumes a crouching posture with bristled hairs, but they found no submissive gesture given by adult bats in this study.

The onset of mating behavior is preceded by an erection (Greenhall 1965; personal observation of two matings at Palo Verde). The male climbs onto the receptive female's back, securely holding her folded wings against her body. The male then transfers his grip to the female's abdominal region and at the same time grasps the nape of her neck with his mouth. During the actual copulations, which lasted 3–4 minutes, I observed pelvic thrusts by the males. I could hear no vocalizations from the pair during either of the matings. At the terminus, the male releases both his forearm and mouth grips, and the female darts off and begins to groom herself. Greenhall (1965) reported one case where other males in the cage clustered around and bit and snapped at the mating pair. This did not occur during either of the matings which I observed in the cave, even though other males and females were within 30 cm of the pair.

Greenhall (1965) and Crespo, Linhart, and Burns (1972) found that *Desmodus* adults spend a fair amount of their time grooming themselves and each other in captivity. Using red-cellophane-covered lights in the Palo Verde cave roost, I frequently noted auto- and allo-grooming. Schmidt and Manske (1973) reported that juvenile *Desmodus* showed no social grooming but received most of the grooming from adult females. In Greenhall's (1965) captive colony, the bats most often hung in two groups of mixed sexes. At both La Pacifica and Palo Verde, roost groups most often consisted of members of both sexes.

Schmidt and Greenhall (1972) have made observations suggesting that dominance was related to sex and size of the individual. In a captive situation with three females and two males, the older and larger male was never disturbed by other bats while feeding. The smaller male was chased away by the older male and two larger females. The two larger females always chased away the smaller, younger female. Occasionally the subordinate female was charged by a feeding male; often she immediately turned and displayed her hindquarters, and the male returned to feed.

Greenhall (1970b) reported that nine bat species are commonly found in roosts with the three species of vampires. During fifteen months I recorded members of the following genera of bats as inhabiting roosts containing *Desmodus*: *Micronycteris, Glossophaga, Carollia, Sturnira, Saccopteryx,* and *Artibeus.* Individuals of these genera were netted along with vampires as they emerged from roosts. With vampires shifting roosts on an opportunistic basis, this might have important consequences on the population of non-vampire bats inhabiting the same tree hollows. However, when I was conducting vampire roost surveys, it was impossible to accurately identify non-vampire bats cohabiting the roost. Thus I only secured data on the total number of vampires and the total number of non-vampire bats inhabiting a particular roost. Looking at these surveys over the fifteen-month study, I tallied what happened to the number of other bats, that is, whether it increased, decreased, or remained the same, when the roost's vampire population increased, decreased, or remained the same.

The data from successive though randomly spaced roost surveys appear in table 27. From this, I note that if the roost's vampire population increases, the tendency for the non-vampire population is to decrease or remain the same but rarely to increase. If the number of vampires decreases, the non-vampire population either increases or remains constant but does not decrease. If the number of vampires remains constant, the number of non-vampire bats may either remain constant or decrease but rarely increases. These data were not collected in a manner allowing statistical analysis. Still the apparent trends may be meaningful. These imply that the vampires may be dominant over other coinhabitants of their roosts, either overtly or indirectly. Almost invariably, when vampires were in a hollow tree roost, they were located at the highest (and darkest) point in the hollow. Bats of other species were located in clusters below the vampires. When vampires were absent from the

TABLE 27

The Interspecific Roosting Habits of Vampires as Determined from Successive Roost Surveys Conducted Randomly over Fourteen Months

Change in the non-vampire population	When the number of vampires increased	When the number of vampires decreased	When the number of vampires did not change
Increased	1	7	1
Decreased	4	0	6
Remained the same	3	5	5

Note: Numbers represent the number of times an event occurred.

roost, either that highest point would be vacant or individuals of another species would have moved into that location. A possible explanation is that natural selection has favored different microhabitat requirements for the different species. But why, then, would a different species move into the vampire's area when there are no vampires and move lower in the tree when vampires return to the roost? Thus, I conclude that vampires are probably dominant over at least some of the bat species coinhabiting their roosts.

Although one instance of attempted "cannibalism" by vampires has been reported (Wimsatt 1959), this occurred in a captive situation, where the author admittedly provided insufficient blood meals. I suspect that the wounds on the bats' bodies were the result of increased aggressiveness due to the conditions of captivity. Perhaps *Desmodus* actively fights with coinhabitants of its roosts; this has never been reported in the literature and was never observed by me. However, I did observe small open wounds on the bodies of vampires occasionally. That a vampire would feed on another bat species or even conspecifics seems extremely unlikely because of the quantity of blood required daily to maintain vital functions.

With a large number of vampires using the same wound, one might expect consequences in the social system and interactions at the feeding site. Indeed, Greenhall, Schmidt, and Lopez-Forment (1971), Schmidt and Greenhall (1972), and Schmidt and van de Flierdt (1973) have discovered dominance hierarchies, possibly combined with territoriality in vampires studied both in the wild and in the laboratory. Greenhall, Schmidt, and Lopez-Forment (1971) have made the only concrete observation of territorial behavior at the cow; one bat, during and *after* feeding from a wound, continually drove invaders away from this wound site. The possibility that vampires are territorial at the feeding site would also be consistent with two additional observations I made:

1) Even though the flight activity patterns for males and females are similar, males apparently do not feed until the early morning (see figure 4 and table 13). This raises the question of what the males are doing when they are out of the roost but not feeding! I have frequently observed vampires darting at fairly high speed above and around a grazing cow herd during a period when the bats would not normally feed on the cows. These observations taken collectively imply that vampires may defend a preferred herd or individual and that males might be engaged in these activities in the earlier part of the evening, before they feed. However, quantitative evidence of territoriality is still lacking, due to technical problems of data collection in the field.

2) As mentioned in chapter 3, I banded a total of 279 vampires on La Pacifica and estimated the population to be in the neighborhood of 100 bats. I also suggested that there are really two populations in the area, one resident and one made up of transient individuals, and that the residents may aggressively exclude the transients from feeding on their cows. This could explain why Schmidt and Greenhall (1972) have concluded that *Desmodus* may exhibit both territoriality and a hierarchical organization at the same time. A dominance hierarchy would permit resident individuals to feed, while territorial behavior would eliminate transient feeders. Of course, this implies that the resident individuals recognize each other (about one hundred acquaintances), and high roost-group turnover would facilitate this. Individual recognition could be based on any of the bat's senses; perhaps even the odor of vampire urine on the cow would be the cue to a form of group recognition. In captivity, a vampire makes every effort not to soil itself or other bats when urinating or defecating (Greenhall 1965).

One might be critical of the hypothesis that vampires are territorial, given the high abundance of food. Normally, a territory is defined as an immobile piece of land defended against conspecifics. Why should a bat waste its energy

PLATE 13. The TeePee frame apparatus attached to a roost tree. Sitting inside and enclosed by nets draped over the frame, the author could capture all bats as they emerged from the roost. It was important that no vampires escaped while roost-group turnover data were being collected.

defending a herd or a particular herd member from other vampires when so many other prey are around and reproductively receptive females are available in the roost? Territorial behavior has been reported in several other bat species (Bradbury 1972; Dwyer 1970), and it may be more common than previously expected. I suggest that if territorial defense does exist in *Desmodus*, it is an evolutionary holdover from the days when large, herding prey species were unavailable. It would make sense to defend a small- to medium-sized prey upon finding it, and even to follow it on successive foraging flights, if prey were hard to locate. Too many vampires feeding from the same, small prey would risk desanguination. A feeding system involving a dominance hierarchy might be evolving currently in response to abundant prey.

Since vampires move among a series of roosts to maintain close proximity to a preferred herd, I had to determine whether the bats move as a social unit or whether there is high turnover on each shift. To assess roost-group fidelity (cohesion), I caught virtually all members of a vampire roost group in the following manner. On two dates (13 June 1972 and 19 January 1973), I attached a teepee frame apparatus to the roost tree. I draped mist nets around this frame, and my assistant enclosed me within the frame with these nets. This allowed capture of vampires as they left the roost and prevented any escapes. With the bats flying out of the roost, I had to be on the inside where I could quickly remove them. I recorded the band numbers of recaptured bats and banded any new captures. On 25 March 1973, a single net was placed directly in front of a different roost with a small entrance. I again captured all bats as they left the roost and recorded the same data as for the first capture method. I calculated an indication of roost-group turnover using the formula:

$$\frac{\text{New associates}}{\text{Total associates}} \times 100.$$

New associates included only those vampires known to be in the vicinity on a previous roost-netting occasion, that is, those banded before the present and previous roost-netting; *total associates* included those known to be in the vicinity previously and new (unbanded) vampires.

The results of three roost samplings were 50, 80, and 100 percent turnover. Thus at a minimum, 50 percent of the roost members were new associates. Therefore, a particular roost group is not always made up of the same individuals. Wimsatt (1969) reported that over one-half of the vampires entering his Mexican cave were new to that cave. Young

(1971) found high roost fidelity and high roost-group fidelity, which he stated in part must be a function of host distribution. Cattle were both numerous and in close proximity to his Costa Rican roosts. Perhaps vampires in Wimsatt's area were following preferred herds, as the vampires on La Pacifica do. The implications for control of genetic inbreeding (suggested by Wimsatt 1969) and rabies virus transmission under the multiple-roost system are obvious.

CHAPTER 7

SEASONAL TRENDS

The tropical dry deciduous forests of Guanacaste Province have two pronounced seasons: the rainy season, from mid-May to mid-November, and the dry season, during the remaining months. At first glance, a rainy season would not appear to influence vampire bat behavior significantly beyond the reduction in flight activity on rainy nights reported by Wimsatt (1969) and Crespo et al. (1972). However, my data on various aspects of vampire biology and behavior indicate otherwise. In chapter 5, I noted the decline in preference for the Brown Swiss cattle breed and adult female bovines during the rainy season. The former was directly related to a change in cattle behavior, distance to nearest neighbor, during the wet season. In this chapter, I will discuss seasonal trends in 1) female reproductive condition, 2) the number of bitten livestock, and 3) the location of bites on the bovine's body.

FEMALE REPRODUCTIVE CONDITION

Wimsatt and Trapido (1952) conducted the first study of the breeding biology and female reproductive cycle in *Desmodus rotundus*, and they concluded that the vampire has no well-defined sexual season and breeds throughout the year. Fleming, Hooper, and Wilson (1972) reached the same conclusion for La Pacifica and Panamanian vampire populations. However, in both instances, conclusions were drawn on the bases that (1) at the same time of the year different females displayed different stages of pregnancy and/or (2) at any time of the year pregnant or lactating females could be netted.

I determined the reproductive condition of each female vampire netted by external palpation and teat manipulation (see Fleming, Hooper, and Wilson 1972 for details). The percentages of females pregnant and/or lactating during each of thirteen months are presented in figure 12. Obviously, the monthly samples were not independent of each other, but I did exclude the second set of data on females captured more than once during any given month. Although it appears that there is an increase in births during the wet season, a chi-square test performed on the numerical data revealed no significant difference from expected values ($\chi_{12}^2 = 12.77$, $P > 0.3$). However, this is not the most powerful statistical test, and from a small population's point of view, an increase in pregnant and/or lactating females from 25 percent during the dry season to 50 percent during the wet season is significant. My sample size for each month was large enough to at least indicate a trend toward wet-season births. In a different region of Costa Rica, Young (1971) concluded that vampires do not breed year-round. Births in his colonies occurred in March and April, just prior to the wet season. Crespo et al. (1961) found maximum vampire births in the Argentinian spring. Goodwin and Greenhall (1961) noted maximum birth periods in Trinidad between April and May and in October–November. R. Burns (personal communication) also found that in his Mexican study area, vampires exhibit a birth period corresponding to the rainy season, and he suggested that the dry season may act as a cue to synchronize breeding in the population.

In a survey of the reproductive cycles of Panamanian mammals, Fleming (1973) concluded that most species produce young when food resources are highest. Sadleir (1969) stated that the period from late pregnancy through birth to lactation is a period when both the mother and infant are susceptible to detrimental effects from the external environment. It is logical to assume that mammals will tend to

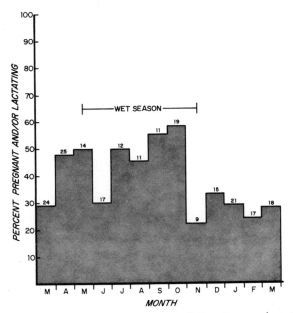

FIG. 12. the percentages of female vampires that were pregnant and/or lactating during each of thirteen months. Sample size for each month is indicated and does not include multiple recaptures during the same month.

produce young when environmental conditions are optimal for survival. Thus, if food resources change through time, either in abundance or accessibility, those individuals breeding so that young are born during a productive period would have the selective advantage. The increased exposure of prey to potential attackers during the wet season, which I have discussed previously, could be advantageous to both mother bats and juveniles learning to feed during this period. The increased energy requirements of pregnant and lactating females should result in decreased selectivity among prey, as predicted by Schoener (1971) and demonstrated for vampires in the present study, and/or increased biting rates, as demonstrated below.

I also looked at reproductive patterns of individual females over a fourteen-month period, and even longer in some cases. The reproductive records for females banded previous to my study were made available by Fleming, Opler, Heithaus, and Wilson (personal communications). These data appear in figure 13. Estimates of length of gestation range from *at least* five months (Wimsatt and Trapido 1952) to seven months (Schmidt and Manske 1973). Since I collected reproductive data by external palpation, I undoubtedly missed small embryos (less than one month) and am unable to state a precise gestation period. Wimsatt and Trapido (1952) also found that at least on occasion, female *Desmodus* exhibit a post-partum estrous. My results are in agreement; I netted four females during the study that were both lactating and pregnant. In figure 13 I note both a female impregnated within six months of prior parturition and females not pregnant for another year. That females apparently do not all exhibit post-partum estrous is curious; it may reflect a change in breeding biology with the advent of domestic cattle. Asdell (1966) has concluded that seasonality, monestrous or polyestrous breeding, and post-partum estrous condition appear to be more related to environmental, nutritional, or climatic factors than to evolu-

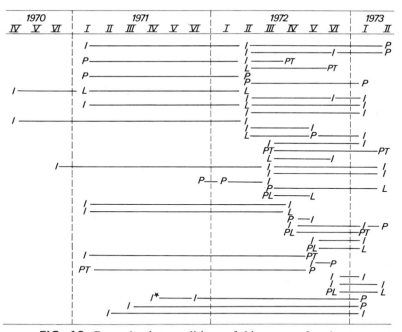

FIG. 13. Reproductive conditions of thirty-seven females captured more than once, during and prior to the study. Roman numerals represent two-month periods listed in chronological order. Key to symbols: *I* = reproductively inactive; *P* = pregnant; *PT* = pregnant-term; *PL* =pregnant and lactating; and *L* = lactating. (*Note*: The reproductive conditions of females caught prior to my study were provided by Fleming, Opler, Heithaus, and Wilson. When first captured, the female marked with a star was an infant bat.)

tionary status. Although one case of twinning in *Desmodus* has been reported (Burns 1970), the vampire typically produces one offspring at each pregnancy (Wimsatt and Trapido 1952).

Finally, in discussing the seasonality of vampire reproduction, I must put forth a hypothesis related to estrous-cow selection and van Tienhoven's comment that bovine hormones might act as birth-control pills on the female vampire (chapter 5). Since the degree of female preference by vampires drops during the wet season, feeding on non-estrous females or males might be the cue that tends to synchronize vampire births each year. Perhaps these hormones cause abortions or result in cessation of estrous cycling, as they do in hormone-injected mice (Asdell 1964). Although I realize this is totally speculative, I am convinced that this hypothesis would provide a fascinating research project for the future.

THE NUMBER OF BITTEN LIVESTOCK

Figure 14 shows the total number of bitten cattle and horses on La Pacifica for each of thirteen months, corrected for changes in the overall cattle and horse population during the same period. For all 1,200 animals, there was a mean of

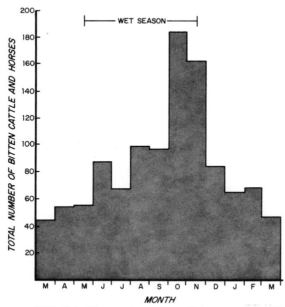

FIG. 14. The total number of bitten cattle and horses during each of thirteen months. Numbers are corrected for changes in the overall cattle and horse population during the same period.

2.8 bitten animals per night ($SD = 1.3$; $SE = 0.4$) over the study period. At first glance, it appears that predation rate increased during the wet season on La Pacifica. Certainly the number of bitten animals per month increased during this period; but this does not necessarily mean that the total amount of blood taken from the domestic stock as a whole increased. Recall that a number of vampires feed from the same wound, and a change in the number of bats per wound could result in the change in the number of wounds (or bitten animals) without an increase in the vampire population. Indeed, Young (1971) found a decrease in the number of vampires per cow during the wet season; the population spreads its attacks over more animals, with fewer multiple feedings from the same wound. This could mean that there are more potential prey available to choose from, and my data on the increase in exposure to potential attackers during the wet season is supportive. Young also reported that vampires spend less time attached to the cow during the wet season. With less evaporative cooling during the windless, wet-season nights, increased blood flow to the skin (Whittow 1962) may result in more profusely bleeding wounds and faster feedings.

To be certain that the lower bite rates during the dry season did not result from vampires selecting wild hosts, stomach contents from bats caught during each season were analyzed. At all times of the year, vampires at La Pacifica fed only on domestic cattle or horses (see table 15).

Recently, predation rates have been published for other study areas. In the state of Puebla, Mexico, Schmidt, Greenhall, and Lopez-Forment (1971) found an average 0.208 bites per animal per night on Zebu stock (the least preferred breed) and 1.984 bites per animal per night on Swiss stock. Thompson, Mitchell, and Burns (1972) reported slightly over one bite per animal per night on three herds in the state of San Luis Potosi, Mexico. In the same area, Linhart, Crespo, and Mitchell (1972) found an average of 0.2 bites per adult bovine and 1.2 bites per calf per night. G. C. Mitchell (personal communication) discovered 54 fresh bites on 12 dairy cows near Managua, Nicaragua, and told me that this was the highest biting incidence he had ever seen. None of the above studies considered a possible seasonal effect. I found a mean 2.8 bitten animals per night out of all 1,200 domestic head, or 0.002 bites per animal per night (see chapter 3). The fact remains that this is the second documented case of a seasonal change in the number of bitten livestock in a given area.

LOCATION OF BITES
ON THE BODY

Although at most times vampires make their feeding wounds on the neck-shoulder region of the bovine, they occasionally bite at odd locations. The percentages of vampire wounds occurring at sites other than the neck-shoulder region of the bovine are given in figure 15, for each of thir-

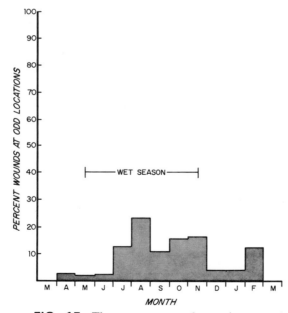

FIG. 15. The percentages of vampire wounds occurring at sites other than the neck-shoulder region of the bovine for each of thirteen months.

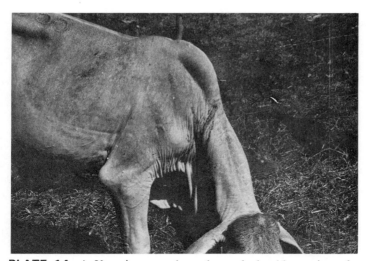

PLATE 14. A Vampire wound on the neck-shoulder region of a bovine. The bats remove only a 3 mm piece of flesh, but due to an anticoagulant substance in their saliva, the wounds bleed freely for several hours.

teen months. A chi-square test for independence revealed a significant difference from expected values ($\chi_{12}^2 = 51.76$, $P < 0.001$); during the wet season there is an increase in the occurrence of wounds at odd locations on the bovine's body.

Several factors could be responsible for this change. Since the distance to nearest neighbor in a herd increases during the rainy season (the bovines are less densely packed), more body regions are accessible. The increase in bites at odd locations could also reflect young bats learning how and what to feed on. After the young vampires reach three months, their mothers feed them blood from their mouths, and at five to six months, juveniles visit prey with their mothers (Schmidt and Manske 1973). Dwyer (1970) has shown that young *Myotis* remain for some time with the mother after weaning, and Möhres (1966) reports that *Rhinolophus* young are guided by the mother on their first flight. Young bats may eventually learn which prey types are easiest to feed upon, which could explain the lower pref-

PLATE 15. A cluster of vampires in a La Pacifica tree roost. Note the banded bats (males on right forearm; females, on left) and the juvenile bat ventro-ventro with a female (to the left of the lowest bat in the photo). The white areas in the picture are fungus growing in the roost. (*Photo courtesy of J. W. Bradbury.*)

erences for females and Swiss during the rainy season. Food preferences of some birds, dogs, and cats may be learned (Kuo 1967). The increased energy requirements of pregnant or lactating females and an increase in the overall population (from births) may result not only in less stringent prey selection but also in less stringent bite-site selection. The possibility of more freely bleeding wounds, due to increased blood flow to the skin during this time, may also reinforce or permit these changes.

Apparently, vampires in different geographical regions inflict their wounds at different locations on the body. In the eastern part of Mexico vampires usually bite burros, horses, and cattle on the cheek and on the neck at the bases of the ears (Dalquest 1955). Goodwin and Greenhall (1961) found that water buffalos in Trinidad and Brazil, which often immerse themselves in water, are frequently bitten inside the nostrils. Greenhall, Schmidt, and Lopez-Forment (1971) stated that most bites on Mexican cows were on the neck or flank. On one Bolivian ranch, vampires were biting exclusively on the cow's back above the pelvic girdle (Mitchell, personal communication). Crespo, Burns, and Linhart (1971) reported that twenty-seven of forty-nine bites inflicted while the prey were standing were on the hoof area, and twenty-one were made at different regions when the bovines were lying down. None of the studies mentioned above took into account the possibility of seasonal changes in vampire biting behavior; however, when considered together, they reiterate the fact that the vampire is an extremely adaptable species, even to the point of selecting different bite sites in different geographical regions and on different prey species.

CHAPTER 8

A COMPARISON
OF VAMPIRE
POPULATIONS

It is always difficult to compare animal populations, since so many variables remain uncontrolled or unexamined. This was particularly true in my study because the prime goal was a thorough understanding of vampire hunting behavior and prey selection at one study site. Thus the comparisons I made were on those factors for which the data were sufficient to warrant comparison or on factors which I found potentially stimulating for future research.

HUNTING SUCCESS

The Palo Verde cave vampires did not appear to use more than the main cave roost during the day. However, within this large limestone outcropping, the bats utilized various crevices at different times. For instance, a cluster of vampires would be in a particular spot on one of my visits; then at the time of the next visit this location would be vacant, but another would be occupied. I thought it would be interesting to compare the foraging success of vampires under this single-roost system with those on La Pacifica, using a series of day-roosts. Table 28 presents the total number of bats captured between 0300 and 0500 hours, grouped by stomach condition and population. A chi-square test for independence revealed a significant difference between La Pacifica and Palo Verde cave vampires ($\chi_2^2 = 6.86$, $P < 0.05$). Vampires were more successful at securing a blood meal on La Pacifica than at Palo Verde. To make this comparison, I am of course assuming that the primary function of vampire flights at both study sites was the same, that of securing a blood meal. I feel this is a safe assumption, given the huge energy cost of flight (McFarland and Wimsatt 1969; McNab 1973).

Due to differences in the ways various authors collected and/or reported foraging efficiency data, I am unable to compare my results with theirs. Wimsatt's (1969) paper on foraging efficiency gives the impression that vampires are highly successful at securing blood meals in Mexi-

TABLE 28

Foraging Efficiency of Vampires at La Pacifica and Palo Verde

	Number of bats caught in each stomach category[a]			
Population	Empty	Undetermined	Full	Percent Full
La Pacifica	21	6	50	65
Palo Verde	54	5	60	50

Note: La Pacifica data were pooled from the entire study; Palo Verde data were pooled from all four samples taken.
[a]Number of bats (males and females) caught between 0300 and 0500 hours.

co. Young (1971) reached the same conclusion but did not offer real data on the proportions of foraging vampires that secured blood meals. La Pacifica vampires have a relatively small foraging range, maintain close proximity to preferred herds by utilizing a series of roosts, and are successful at finding suitable prey under these circumstances. At Palo Verde, cattle are not maintained in distinct herds, and they roam freely through semi-open forest and scrub vegetation. I suspect a general difficulty in locating prey under these conditions. If food is less accessible (effectively, less abundant), all foraging models summarized by Schoener (1971) would predict less selectivity by vampires among the various potential prey. From immunological analysis on stomach contents of Palo Verde bats, I know the bats were feeding only on domestic stock; however, I was unable to determine host preferences among the domestics.

SEX RATIOS AND REPRODUCTIVE CONDITION

I have already demonstrated a difference in foraging behavior and success between male and female vampires. To make the pooling of foraging data from the two La Pacifica subpopulations and the comparison between the La Pacifica and Palo Verde populations more valid, I checked the sex ratio of each population against an expected ratio of one to one. Treating each month as a sample point, the number of males and females captured in each population are presented in table 29. Chi-square tests revealed no significant differences from an expected male:female ratio of one to one for the two La Pacifica subpopulations ($\chi_{12}^2 =$ 14.35 and 6.13, $P > 0.3$) but did show a significant departure for the Palo Verde cave population ($\chi_4^2 = 35.65$, $P < 0.001$). There are statistically more females than males inhabiting the cave.

One can extract sex ratios from data in the papers of Wimsatt (1969) and L.-Forment, Schmidt, and Greenhall

TABLE 29

Number of Males and Females Caught in Each Vampire Population

Population	Sample month[a]														Total
	Mar.	Apr.	May	June	July	Aug.	Sept.	Oct.	Nov.	Dec.	Jan.	Feb.	Mar.	Apr.	
La Pacifica															
Rio Corobici															
Males	3	12	1	11	7	11	10	14	20	14	13	18	18	—	152
Females	6	8	1	15	2	5	5	6	3	11	17	9	17	—	105
Rio Tenorito															
Males	11	13	2	4	11	2	6	1	6	8	4	5	2	—	75
Females	17	15	11	1	10	6	5	7	4	3	5	8	1	—	93
Palo Verde															
Males	—	—	—	—	—	—	—	—	—	—	62	15	29	52	158
Females	—	—	—	—	—	—	—	—	—	—	139	52	60	91	342

[a]For La Pacifica subpopulations, each month is treated as a sample point; for Palo Verde, each visit is treated as a sample point.

(1971). In their vampire populations, numbers of males and females were about equal. The higher proportion of females in the Palo Verde population is suggestive of a maternity or nursery colony system, where pregnant and/or lactating females congregate to raise young. Crespo et al. (1961) reported sexual segregation for Argentinian vampires, and R. Burns (personal communication) found that sexual segregation *may* occur in some Mexican roosts during the birth season. It is interesting that not all populations segregate into nursery colonies; on La Pacifica I noted new-born, suckling infants with their mothers in normal roosts. The use of a maternity roost system probably depends on the availability of a suitable roost in the area; on La Pacifica there was no such suitable roost, since all known roosts were in relatively small tree hollows or rock-cliff holes.

There was at least a trend toward a higher proportion of pregnant and/or lactating females at Palo Verde than at La Pacifica. The numbers of pregnant and/or lactating females and inactive females caught during the same time of year for La Pacifica and Palo Verde are presented in table 30. A chi-square test for independence revealed no significant difference at the $P < 0.05$ level but one at the

TABLE 30
Female Vampire Reproductive Condition
on La Pacifica and on Palo Verde

Population	Total number of females	
	Pregnant/lactating	Inactive
La Pacifica	10	28
Palo Verde	66	86

Note: Data are from February to May 1973 only.

$P < 0.10$ level ($x_2^2 = 3.02$). Palo Verde females may have a higher reproductive rate than those on La Pacifica; or the population's breeding season may be out of synchronization with that of La Pacifica, but this is doubtful, given the similar rainfall patterns for the two areas.

CHAPTER 9

VAMPIRE BAT
CONTROL

In this chapter, I will discuss three points of a more applied nature: (1) the vampire bat as a disease vector; (2) established methods for controlling vampire populations; and (3) recommendations for vampire control, based on what is now known about vampire behavior and ecology. As such, this chapter will be of greater interest to public and animal health researchers and agriculturalists in Central and South America. However, the vertebrate zoologist might also find it useful as an example of how one can use basic knowledge of an animal's behavior and ecology in solving health and economic problems.

THE VAMPIRE AS A DISEASE VECTOR

Vampire bats transmit equine and bovine trypanosomiasis (Hoare 1965), Venezuelan equine encephalomyelitis (Rosenthal 1972), and, of course, paralytic rabies virus. Vampire-borne rabies has been declared the primary livestock disease problem of Latin America and a formidable obstacle to the expansion of its agricultural economy (World Health Organization Expert Committee on Rabies 1966). Steele (1966) found that more than one million animals were lost annually, with an associated financial loss exceeding 100 million dollars. Man, cattle, horses, sheep, goats, and swine are susceptible to vampire-borne rabies, and *Desmodus* is considered the principle vector of this disease among livestock (Pawan 1936; Acha 1968; Schmidt, Greenhall, and Lopez-Forment 1970; and Greenhall 1972*b*).

We know little about the course of rabies infection in the vampire itself. Normally for other species the virus is fatal. However, Pawan (1936) and Torres and Queiroz Lima (1936) demonstrated that many vampires survive the disease. The virus appears in the vampire's saliva two weeks after infection and can be detected there for up to three months. Some bats may die, while others survive. Sugay and Nilsson (1966) and Williams (1960) have reported infection rates of about 1 percent in brain and saliva gland tissues of clinically normal vampires. Constantine (1971) has provided an excellent review article on bat rabies in which he concludes that infection in vampires appears to be a closed cycle, with the exception of viral exchange with other bat species, and that man and livestock are infected tangentially.

Ruiz-Martinez (1963) has concluded that endemic zones are characterized by rabies outbreaks every two or three years. In 1968, Costa Rica experienced such an increase in the number of cases reported (Centro Panamericano de Zoonosis 1972), and it could be on the verge of another upswing at the time of this writing. Unfortunately, most rabies investigations involve counts of infected cattle and not infection rates within the principle vector populations of vampires. Dr. R. A. Vargas, Chief of the Rabies

Section, Costa Rica Ministry of Public Health, has informed me of unpublished evidence that bovine rabies appears seasonally, peaking during the wet season. Of six vampires from the La Pacifica area tested for rabies during the rainy season, three were positive. Of thirty bats tested during the dry season, none tested positively.

What factors in the vampire's behavior and ecology make it particularly important as a rabies vector? Pawan (1936) showed that *Desmodus* is capable of transmitting the virus to other animals and conspecifics by its bites. Thus the aggressiveness of vampires noted in chapter 6 and the wounds observed on vampire bodies by Wimsatt (1959) and myself could have direct implications for rabies transmission. Vampires roost with other bat species, and the disease might spill over into these species by a number of mechanisms. Vampires switch roosts fairly often, and there is a high degree of mixing with new, conspecific associates as a result of these shifts, which could result in higher transmission rates within the vampire population than otherwise expected. The high degree of mixing with unfamiliar animals might also affect the level of aggression (biting) within the colony and, in turn, virus transmission.

I am aware of no published data confirming wet-season peaks in bovine rabies, but I have heard evidence of such peaks from a number of investigators. To confirm this and determine its cause, future researchers must consider infection rates both in the cattle and in the vampire populations, as well as seasonal behavior changes in the vampire. Where vampires breed more or less seasonally, one might expect a seasonal difference in the level of intraspecific aggression, which could affect transmission rates between vampires and which could in turn affect infection rates in the prey. On the other hand, a higher wet-season death rate among cattle does not necessarily mean an increased rabies infection rate among vampires. Recall that vampires spread their attacks over more animals (fewer bats using the same cow and wound) during the wet season. Given this feeding system, the same proportion of infected bats would have a much greater impact on the prey population during the rainy season that at other times. Therefore, all three factors—infection rate in prey species, infection rate in the predator-vector species, and seasonal changes in the behavior of the vector—must be considered in future studies of vampire-borne rabies.

CONTROL METHODS

Trinidad was the first country to establish a government program to control vampire bats (Greenhall 1970*b*). From the mid-1930s to the present decade, various control methods have been developed and tested, and these are described in detail by Greenhall (1970*b*). Therefore, I will only acquaint the reader with these and then move on to the large-scale chemical control methods developed in the last two or three years and being implemented in various nations today.

All of the earlier control methods were designed to either destroy or disturb the bats or to protect potential prey either from attacks or rabies. Dynamite, poison gas, smoke and fire, firearms, screens to seal off roosts, bat traps, and various types of capture nets have been used to destroy the bats in the roost. Obvious disadvantages are general destruction of the habitat and nonselectivity of the control agent. Recall that many economically important bat species (insect eaters and plant pollinators) live in the same roosts with the vampires. Even before the disposal of vampires by means of trap- or net-capture techniques, individuals must be trained in the proper identification of bat species. Greenhall (1963) developed a strychnine poisoning technique, in which a strychnine syrup is applied to fresh vampire wounds, and bats returning to feed from a treated wound are selectively killed. However, this is impractical for a large-scale cattle operation.

With respect to protecting potential prey either from attacks or rabies, livestock and people may be housed in enclosures with protective screening over openings. This is perhaps practical for chickens and for humans with high-quality housing but certainly not for a large cattle herd. Besides, the vampire is quite adept at finding and passing through small openings (Greenhall 1970b). Cattle and other livestock may be vaccinated against rabies, and humans may receive postexposure treatment, but the problem persists.

Where vampires have a relatively small foraging range concentrated along rivers, I suggest a new method of control to be tested. The four nights around calendar full moon should be a stressful period for the vampire population due to lack of food. For either the four days prior to the calendar full moon or just after that period, cattle herds could be moved to distant pastures away from rivers or known roots and grazed there. The available foraging time during either period may not be sufficient to allow successful herd location, which could produce a stress on the vampire population significant enough to cause death by starvation and would only affect the vampire species. Although *Desmodus* is an adaptable species, I doubt that it would alter its search patterns and prey preferences, which may be learned, and attack humans within a four-day period.

In 1972, the U.S. Agency for International Development, in cooperation with the U.S. Department of the Interior and the Instituto Nacional de Investigaciones Pecuarias, Mexico, announced the development of new chemical control techniques of reducing vampire populations (Linhart, Crespo, and Mitchell 1972; Thompson, Mitchell, and Burns 1972; Mitchell and Burns 1973). These techniques rely on the use of the anticoagulant diphenadione (2-diphenylacetyl-1, 3-indandione) applied either as a petroleum-jelly paste mixed with the anticoagulant smeared onto the back of a captured vampire or as a thick liquid form of the chemical injected into the rumen compartment of the bovine stomach. A second anticoagulant, chlorophacinone

(2-[(*p*-chlorophenyl) phenylacetyl]-1, 3-indandione), has also been successfully tested for topical application on vampires (Linhart, Crespo, and Mitchell 1972).

The topical application methods make direct use of behavioral observations on vampire bats. The bats spend much time grooming themselves *and each other* in the roost. After the chemical is applied, the bat returns to the roost where it and conspecifics ingest the anticoagulant through grooming. Spontaneous hemorrhaging occurs in the digestive tract, and in the presence of the anticoagulant, the bat dies of internal bleeding. With the intrarumenal-injection technique, the dosage is set so as to be harmless to the bovine, but it is high enough to kill a vampire feeding on blood in which the anticoagulant is circulating (Thompson, Mitchell, and Burns 1972; Mitchell and Burns 1973). Both methods have been shown to reduce the local bite incidence by over 90 percent for over six months.

Mitchell and Burns (1973) have discussed the advantages and disadvantages of each control method. Topical application on the bats can be used around all domestic species, whereas only bovines should be injected. Treating captured bats is more effective for fast, complete control, and in a rabies outbreak, this is essential. However, topical application requires special equipment (mist nets, lights, heavy gloves) and specially trained control teams, the members of which have been vaccinated against rabies, for proper bat identification. The cattle-injection method eliminates direct contact with all species of bats and is definitely species-specific; however, it requires cattle chute facilities and is more expensive and time consuming than treating the bats.

RECOMMENDATIONS FOR CONTROL

The International Union for the Conservation of Nature and Natural Resources (1968) passed a resolution recommending that the necessary control of vampire bats be based on sound biological and ecological studies, since these bats often roost with other species. In the interest of both bat conservation and efficient chemical control of vampires as promoted by the U.S. Agency for International Development, I make the following recommendations based on my findings in Costa Rica:

1) A chemical control program against vampire bats should *not* be instigated at this time in Costa Rica. Predation rates and rabies incidence (Costa Rica Ministry of Public Health, personal communication) are too low to warrant a control program. In time, however, this situation might change, and the next recommendation should be followed.

2) In any nation where vampire control (chemical) programs are being considered, a thorough investigation of rabies incidence (paralytic bovine type), vampire population densities, and domestic animal densities should be conducted to indicate specific geographical problem areas. There, and only there, should chemical control be instigated. In recommending when treatment should be considered, Mitchell and Burns (1973) state that one fresh bite generally represents one vampire; as discussed in the preceding chapters, this is not the case, given multiple feedings from the same wounds. Therefore, actual population density estimates should be utilized in making this decision.

3) When the situation warrants chemical control, I recommend systemic treatment of livestock rather than topical application on the bats. As of this date, no reports have been published showing data on the effects of the anticoagulant on other species of bats, and Linhart, Crespo, and Mitchell (1972) themselves recommend further studies on this. They are correct when they state that vampires using a roost containing other bat species cluster together in different "niches." However, my observations indicate that other species are often located directly below the vampire clusters; this is potentially dangerous if the vampire urine contains the ingested chemical. Additionally, the intrarumenal

injection involves neither the proper identification of vampires—and I question whether local control teams would bother to free non-vampire species from the nets—nor the potentially dangerous handling of rabid bats. Initial laboratory tests have shown no harmful effects to bovines (Thompson, Mitchell, and Burns 1972).

4) Based specifically on my findings and the need to keep control costs minimal, when the situation demands chemical control of vampires, I recommend systemic treatment in the following manner: a) livestock (bovines) should be treated at the end of the dry season or early in the wet season, since this is a period of high multiple feedings (bats/ cow) and just prior to maximum vampire births; (b) livestock should not be injected during the full-moon phase, since this is a period when most vampires do not feed (also discussed by Mitchell and Burns 1973); and (c) to insure highest benefit for cost, heavily bitten calves, cows, and breeds (European) should be injected before nonpreferred stock is treated.

CHAPTER 10

CONCLUDING
REMARKS

I would like now to conclude with a short discussion of the predator-prey relationships I have found, the future of La Pacifica vampires, and the most exciting prospects for future research. The major relationships discovered during this field project can be summarized as in figure 16. I found exposure to potential vampire attacks to be the critical factor in the predator-prey relationship. Calf preference may be explained on the basis of exposure. Calves remain inactive during the night while their mothers graze the pasture, and when mothers and calves sleep at night, they do so with greater distances between individuals than do members in an all-adult herd. Estrous cows are located on the edge of a sleeping herd, where most potential prey are found. Adult female preference is probably related to exposure *through* the estrous cow preference, since most of the females bitten

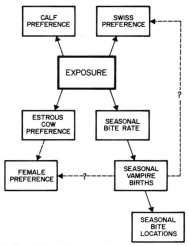

FIG. 16. The major relationships discovered during the field project.

were probably in peak estrous. The Swiss breed preference may also be explained on the basis of exposure. Swiss animals sleep on the perimeter of a tightly packed herd. Exposure increases during the rainy season, since greater distances to nearest, sleeping neighbors are maintained. With a greater number of exposed potential prey, the total number of bitten animals increases. Effectively, the vampires spread their bites around; this was confirmed by Young (1971) when he observed fewer vampires per cow during the rainy season. Most tropical mammals have birth seasons when food supplies are most abundant. Vampires exhibit increased births during the wet season, when more prey are exposed to potential attacks. The decline in Swiss breed- and adult female-preferences during the rainy season may be the result of increased energy demands on pregnant and/or lactating mothers who, in theory at least, should be less selective in diet, or it may result from young vampires learning which prey types are easiest targets. Lastly, seasonal trends in the location of wounds on the bovine's body may be the result of greater exposure of the entire bovine body (greater distances between individuals) and/or young bats again learning how and where to secure their blood meals.

When I left La Pacifica, the owner was considering various changes in his management program. Among the possibilities were (1) a decrease in breeding stock (mostly females and Brown Swiss) and an increase in beef stock (males) and (2) a general decrease in the proportion of Swiss stock and an increase in Brahma stock. At first glance, this would seem detrimental to the vampire population that prefers cows and the Swiss breed. But the vampire seems quite adaptable, and these changes would probably result in increased predation on males and Brahma stock or increased multiple feedings on the remaining females and Swiss stock. The latter could be detrimental to the few individuals attacked. Second, based on the low bite rate and low vampire population estimates relative to other Latin nations, I would feel safe in predicting an increase in the vampire population over the next few years, assuming no decrease in the number of available roosting sites. At present, the hacienda owner does not plan further land clearing.

It is difficult to determine which environmental factor is limiting the vampire population today. Theoretical models predicting high selectivity when food is common indicate these bats are not limited by food, since they exhibit high selectivity. Dalquest (1955) believed the populations were regulated locally by the presence of suitable roosts. Yet I found that vampires regularly shift roosts, leaving "suitable" roosts unoccupied for days. However, a number of roosts may actually be required so that the bats can maintain close proximity to preferred hosts. Rabies, perhaps appearing on a cyclical basis, might also limit the population directly, by causing vampire deaths, or indirectly, by elimination of prey. Vampires were infected with rabies virus prior to the discovery of America (Constantine 1971), and if the disease co-evolved with this species, it could still perform regulatory functions.

Fortunately, total eradication of vampire bats seems an impossibility because of their wide distribution, abundance, adaptability, and often inaccessible roosts (Greenhall 1972*b*), and the possibility of future research remains. Considering modern ecological theory and interest and the findings I have made, I see two pertinent avenues of future research. First, the possible interaction between prey selection (estrous cows and calves), bovine reproductive hormones, and reproduction in female vampires should be of interest to physiologists and population ecologists as a potential population-regulating mechanism. Second, the interaction between prey distribution, roosting systems, and foraging efficiency could provide an excellent model to test future theories on foraging strategies, since the distribution of food resources can be manipulated, the distribution of roosts can be determined, and the foraging success of the vampires can be measured. I have speculated on these and other hypotheses throughout this text, and I believe that many interesting and important discoveries lie just around the corner. Hopefully, the present research will serve as an adequate stimulus for future studies on the common vampire bat, *Desmodus rotundus*.

REFERENCES

Acha, Pedro N. 1968. Epidemiologia de la rabia bovina paralitica transmitida por los quiropteros. *Bol. Of. Sanit. Panam.* 64(5): 411–30.

Agricultúra de las Amerícas. 1972. Derriengue o rabia paralitica: lo que ud. puede hacer contra el vampiro. *Año* 21(4): 16–36.

Arata, A. A.; Vaughan, J. B.; and Thomas, M. E. 1967. Food habits of certain Columbian bats. *J. Mammal.* 48(4): 653–55.

Asdell, S. A. 1964. *Patterns of mammalian reproduction.* 2d ed. Ithaca: Cornell University Press.

———. 1966. Evolutionary trends in physiology of reproduction. In *Comparative biology of reproduction in mammals*, ed. I. W. Rowlands, pp. 1–13. London: Academic Press.

Baenziger, H. 1968. Preliminary observations on a skin-piercing blood-sucking moth (*Calyptra eustrigata* [Hmps.] [Lep., Noctuidae]) in Malaya. *Bull. Ent. Res.* 58(1): 159–63.

Balch, C. C. 1955. Sleep in ruminants. *Nature* 175: 940–41.

Benzoni, M. Girolame. 1565; reprint 1967. *La Historia del Mundo Nuevo.* Excerpts were translated for me by Benjamin Waite. Caracas: Biblioteca de la Academia Nacional de la Historia.

Bonaccorso, Frank J., and Turner, Dennis C. 1971. The vertical stratification of bats in neotropical forests. Abstract read at Second Symposium on Bat Research, University of New Mexico, Albuquerque.

Bradbury, Jack W. 1972. Tuning in on the bat. In *The marvels of animal behavior*, ed. Thomas B. Allen, pp. 112–25. Washington, D.C.: National Geographic Society.

Brakel, W. J.; Rife, D. C.; and Salisbury, S. M. 1952. Factors associated with the duration of gestation in dairy cattle. *J. Dairy Sci.* 35(3): 179–94.

Brown, James H. 1968. Activity patterns of some neotropical bats. *J. Mammal.* 49(4): 754–57.

Brownlee, A. 1950. Studies on the behavior of domestic cattle in Britain. *Bull. Anim. Behav.* 8: 11–20.

Buettiker, W. 1959. Observations on the feeding habits of adult Westermanniinae (Lepid., Noctuidae) in Cambodia. *Acta Trop. (Basel)* 16(4): 356–61.

Burns, Richard J. 1970. Twin vampire bats born in captivity. *J. Mammal.* 51(2): 391–92.

Centro Panamericano de Zoonosis. 1972. Rabies surveillance for the Americas. *Vigilancia Epidemiologica* 4(8): 1–6.

Chase, Julia. 1972. Role of vision in echolocating bats. Ph.D. diss., Indiana University.

Constantine, Denny G. 1971. Bat rabies: current knowledge and future research. In *Rabies*, ed. Yasuiti Nagano and Fred Davenport, pp. 253–65. Baltimore: University Park Press.

Crespo, J. A.; Vallena, J. M.; Blood, B. D.; and de Carlo, J. M. 1961. Observaciones ecologicas del vampiro *Desmodus r. rotundus* (Geoffroy) en el norte de Cordoba. Revista Museo Argentino de Ciencias Nat. "Bernardino Rivadava." *Ciencias Zool.* 6: 131–60.

Crespo, Raul Flores; Burns, Richard J.; and Linhart, Samuel B. 1971. Comportamiento del vampiro (*Desmodus rotundus*) durente su alimentacion en ganado bovino en cautiverio. *Tecnica Pecuaria en Mexico* 18: 40–44.

Crespo, Raul Flores; Linhart, Samuel B.; and Burns, Richard J. 1972. Compartamiento del vampiro (*Desmodus rotundus*) en cautiverio. *Southwest. Nat.* 17(2): 139–43.

Crespo, Raul Flores; Linhart, Samuel B.; Burns, Richard J.; and Mitchell, G. Clay. 1972. Foraging behavior of the common vampire bat related to moonlight. *J. Mammal.* 53(2): 366–68.

Dalquest, Walter W. 1955. Natural history of the vampire bats of eastern Mexico. *Amer. Midl. Nat.* 53(1): 79–87.

Darwin, C. 1890. *Naturalist's voyage around the world*. London: John Murray.

de Azara, F. 1935. Memoria sobre el estado rural del Rio de la Plata en 1801. In *Felix de Azara, Siglo XVIII*, ed. E. Alvarex Lopez, pp. 230–51. Madrid: Aguilar.

De Oviedo y Valdes, F. 1526; reprint 1950. *Sumario de la natural historia de las Indias.* Mexico: Fondo de Cultura Economica.

DiSanto, P. E. 1960. Anatomy and histochemistry of the salivary glands of the vampire bat, *Desmodus rotundus murinus. J. Morphol.* 106: 301–35.

Ditmars, Raymond L., and Greenhall, Arthur M. 1935. The vampire bat: a presentation of undescribed habits and review of its history. *Zoologica* 19(2): 53–76.

Dixon, A. F. G. 1959. An experimental study of the searching behavior of the predatory coccinellid beetle *Adalia decempunctata* (L). *J. Anim. Ecol.* 28: 259–81.

Dwyer, P. D. 1970. Social organization in the bat *Myotis adversus. Science* 168: 1006–8.

Emlen, J. Merritt. 1966. The role of time and energy in food preference. *Amer. Nat.* 100(916): 611–17.

Fleming, Theodore H. 1973. The reproductive cycles of three species of opossums and other mammals in the Panama Canal Zone. *J. Mammal.* 54(2): 439–55.

Fleming, Theodore H.; Hooper, Emmet T.; and Wilson, Don E. 1972. Three Central American bat communities: structure, reproductive cycles, and movement patterns. *Ecology* 53(4): 555–69.

Foster, Mercedes S. 1969. Synchronized life cycles in the orange-crowned warbler and its mallophagan parasites. *Ecology* 50(2): 315–23.

Goodwin, G. G., and Greenhall, A. M. 1961. A review of the bats of Trinidad and Tobago. *Bull. Amer. Mus. Nat. Hist.* 122: 187–302.

Goss-Custard, J. D. 1970. Feeding dispersion in some overwintering wading birds. In *Social behavior in birds and mammals*, ed. J. H. Crook, pp. 1–35. New York: Academic Press.

Gould, Edwin. 1970. Echolocation and communication in bats. In *About bats*, ed. Bob H. Slaughter and Dan W. Walton, pp. 144–61. Dallas: Southern Methodist University Press.

Gould, Edwin; Woolf, Nigel Keith; and Turner, Dennis Clair. 1973. Double-note communication calls in bats: occurrence in three families. *J. Mammal.* 54(4): 998–1001.

Greenhall, Arthur M. 1963. Use of mist nets and strychnine for vampire control in Trinidad. *J. Mammal.* 44(3): 396–99.

——. 1965. Notes on the behavior of captive vampire bats. *Mamalia* 29(4): 441–51.

——. 1968. Problems and ecological implications in the control of vampire bats. *International Union for the Conservation of Nature Publications, New Series* (Morges), 13: 94–102.

——. 1970*a*. The use of a precipitin test to determine host preferences of the vampire bats, *Desmodus rotundus* and *Diaemus youngi. Bijdragen tot de Dierkunde* 40(1): 36–39.

——. 1970*b*. Vampire bat control: a review and proposed research program for Latin America. *Proceedings 4th Vertebrate Pest Control Conference*, ed. California Vertebrate Pest Committee (Davis: University of California Press), pp. 41–54.

——. 1972*a*. The biting and feeding habits of the vampire bat, *Desmodus rotundus. J. Zool., London* 168: 451–61.

——. 1972*b*. The problem of bat rabies, migratory bats, livestock and wildlife. Transactions of 37th North American Wildlife and Natural Resources Conference, 12–15 March. Washington, D.C.: Wildlife Management Institute.

Greenhall, Arthur M., and Paradiso, John L. 1968. *Bats and bat banding*. U.S. Dept. of Interior, Bureau of Sport Fisheries and Wildlife, Resource Publication No. 72. Washington, D.C.: U.S. Government Printing Office.

Greenhall, Arthur M.; Schmidt, Uwe; and Lopez-Forment, William. 1969. Field observations on the mode of attack of the vampire bat (*Desmodus rotundus*) in Mexico. *An. Inst. Biol. Univ. Nal. Auton. Mexico Ser. Zool.* 2: 245–52.

——. 1971. Attacking behavior of the vampire bat, *Desmodus rotundus*, under field conditions in Mexico. *Biotropica* 3(2): 136–41.

Hafez, E. S. B.; Schein, M. W.; and Ewbank, R. 1969. In *The behavior of domestic animals*, ed. E. S. B. Hafez, pp. 236–95. 2d ed. Baltimore: Williams and Wilkins.

Hansel, W., and Echternkamp, S. E. 1972. Control of ovarian function in domestic animals. *Amer. Zool.* 12: 225–43.

Hawkey, C. M. 1966. Plasminogen activator in saliva of the vampire bat, *Desmodus rotundus. Nature* 211: 434–35.

——. 1967. Inhibitor of platelet aggregation present in saliva of the vampire bat *Desmodus rotundus. Br. J. Haematol.* 13: 1014–20.

Heithaus, E. Raymond; Fleming, Theodore H.; and Opler, Paul A. Foraging patterns and resource utilization by eight species of bats in a seasonal tropical forest. *Ecology* (in press).

Herschkovitz, P. 1969. Recent mammals of the Neotropical region: a zoogeographic and ecological review. *Q. Rev. Biol.* 44(1): 1–70.

Hoare, Cecil A. 1965. Vampire bats as vectors and hosts of equine and bovine Trypanosomes. *Acta Trop. (Basel)* 22(3): 204–16.

Holdridge, L. R. 1967. *Life zone ecology*. San Jose, Costa Rica: Tropical Science Center.

International Union for the Conservation of Nature and Natural Resources. 1968. Resolution No. 1, vampire bats. *International Union for the Conservation of Nature Publications, New Series* (Morges), 13:102.

Jahoda, John C. 1973. The effect of the lunar cycle on the activity pattern of *Onychomys leucogaster breviauritus*. *J. Mammal.* 54(2): 544–49.

Johnston, James E. 1963. Response to environment. In *Cross-breeding beef cattle*, ed. T. J. Cunha, M. Koger, and A. C. Warnick, pp. 61–67. Gainesville: University of Florida Press.

Jolly, G. M. 1965. Explicit estimates from capture-recapture data with both death and immigration-stochastic model. *Biometrika* 52(1, 2): 225–47.

Kavanau, J. Lee, and Ramos, Judith. 1972. Twilights and onset and cessation of carnivore activity. *J. Wildl. Mgmt.* 36(2): 653–57.

Kilgour, R., and Scott, T. H. 1959. Leadership in a herd of dairy cows. *Proc. N. Z. Soc. Anim. Prod.* 19: 36–43.

Koopman, Karl F., and Jones, J. Knox, Jr. 1970. Classification of bats. In *About bats*, ed. Bob H. Slaughter and Dan W. Walton, pp. 22–29. Dallas: Southern Methodist University Press.

Kuo, Zing-Yang. 1967. *The dynamics of behavior development: an epigenetic view*. New York: Random House.

Laing, J. 1938. Host finding by insect parasites. II. The chance of *Trichogramma evanescens* finding its hosts. *J. Exp. Biol.* 51: 281–302.

Lampkin, G. H., and Quarterman, J. 1962. Observations on the grazing habits of grade and zebu cattle, II. Their behavior under favourable conditions in the tropics. *J. Agric. Sci.* 59(1): 119–23.

Layne, J. N., and Benton, A. H. 1954. Speeds of small mammals. *J. Mammal.* 35: 103–4.

L.-Forment, William; Schmidt, Uwe; and Greenhall, Arthur M. 1971. Movement and population studies of the vampire bat (*Desmodus rotundus*) in Mexico. *J. Mammal.* 52(1): 227–28.

Linhart, Samuel B. 1973. Age determination and occurrence of incremental growth lines in the dental cementum of the common vampire bat (*Desmodus rotundus*). *J. Mammal.* 54(2): 493–96.

Linhart, Samuel B.; Crespo, Raul Flores; and Mitchell, G. Clay. 1972. Control of vampire bats by topical application of an anticoagulant, chlorophacinone. *Bol. Of. Sanit. Panam.* 6(2): 31–38.

Lloyd, M. 1967. Mean crowding. *J. Anim. Ecol.* 36: 1–30.

MacArthur, Robert H. 1972. *Geograhical ecology: patterns in the distribution of species.* New York: Harper and Row.

MacArthur, Robert H., and Pianka, Eric. 1966. An optimal use of a patchy environment. *Amer. Nat.* 100(916): 603–9.

McFarland, William N., and Wimsatt, William A. 1969. Renal function and its relation to the ecology of the vampire bat, *Desmodus rotundus. Comp. Biochem. Physiol.* 28: 985–1006.

McNab, Brian K. 1969. The economics of temperature regulation in neotropical bats. *Comp. Biochem. Physiol.* 31: 227–68.

——. 1973. Energetics and the distribution of vampires. *J. Mammal.* 54(1): 131–44.

Mann, G. 1951. Biologia del vampiro. *Biologica (Santiago)* 12–13: 3–24.

——. 1960. Neurobiologia de *Desmodus rotundus. Invest. Zool. Chil.* 6: 79–99.

Mead-Briggs, A. R., and Rudge, A. J. B. 1960. Breeding of the rabbit flea, *Spilopsyllus cuniculi* (Dale): requirement of a 'factor' from a pregnant rabbit for ovarian maturation. *Nature* 187: 1136–37.

Mech, L. D. 1966. The wolves of Isle Royale. *Fauna of National Parks of U.S, Fauna Ser.* 7: 1–210.

Mitchell, G. Clay, and Burns, Richard J. 1973. *Chemical control of vampire bats.* Denver: U.S. Bureau of Sport Fisheries and Wildlife.

Mitchell, G. Clay, and Tigner, James R. 1970. The route of ingested blood in the vampire bat (*Desmodus rotundus*). *J. Mammal.* 51(4): 814–17.

Möhres, F. P. 1966. Communicative characters of sonar signals in bats. In *Animal sonar systems, biology and bionics, tome II*, ed. R. G. Bushnel, pp. 939–45. Paris: Laboratoire de Physiologie Acoustique.

Moran, J. B. 1970. Brahman cattle in a temperate environment. II. Adaptability and grazing behavior. *J. Agric. Sci. (Camb.)* 74(2): 323–27.

Pawan, J. L. 1936. Rabies in the vampire bat of Trinidad, with special reference to the clinical course and the latency of infection. *Ann. Trop. Med. Parasitol.* 30: 401–22.

Payne, W. J. A.; Laing, W. I.; and Raivoka, E. N. 1951. Grazing behavior of dairy cattle in the tropics. *Nature* 167: 610–11.

Plasse, D.; Warnick, A. C.; and Koger, M. 1970. Reproductive behavior of *Bos indicus* females in a subtropical environment. IV. Length of estrous cycle, duration of estrous, time of ovulation, fertilization and embryo survival in grade Brahman heifers. *J. Anim. Sci.* 30(1): 63–72.

Pritchard, G. 1965. Prey capture by dragonfly larvae (Odonata: Anisoptera). *Can. J. Zool.* 43: 271–89.

Rhoad, A. O. 1938. Some observations on the response of purebred *Bos taurus* and *Bos indicus* cattle and their crossbreed types to certain conditions of the environment. *Proceedings of the American Society of Animal Production, 31st Annual Meeting*, pp. 284–95.

Romaña, C. 1939. Action anticoagulante de la salive du vampire *Desmodus rotundus rotundus* (Geof.). *Bull. Soc. Pathol. Exot.* 32: 339–403.

Romer, Alfred S. 1959. *The vertebrate story.* Chicago: University of Chicago Press.

Rosenthal, Jerry E. 1972. AID finds ways to control vampire bat. *War on Hunger* 6(6): 11–23.

Rothschild, Miriam. 1965. The rabbit flea and hormones. *Endeavour* 24(93): 162–68.

Ruiz-Martínez, C. 1963. Epizootologia y profilaxis regional de la rabia paralitica en las Americas. *Rev. Veter. Venezolana* 14(79): 71–173.

Sadleir, R. M. F. S. 1969. *The ecology of reproduction in wild and domestic mammals.* London: Methuen.

Schein, M. W., and Fohrman, M. H. 1955. Social dominance relationships in a herd of dairy cattle. *Brit. J. Anim. Behav.* 3: 45–55.

Schmidt, U. 1972. Die sozialen Laute juveniler Vampirfledermäuse (Desmodus rotundus) und ihrer Mütter. *Bonn. Zool. Beitr.* 23: 310–16.

———. 1973. Olfactory threshhold and odor discrimination of the vampire bat (*Desmodus rotundus*). *Period. Biol.* 75: 89–92.

Schmidt, U., and Greenhall, A. M. 1971. Untersuchungen zur geruchlichen Orientierung der Vampirfledermäuse (*Desmodus rotundus*). *Z. vergl. Physiologie* 74: 217–26.

——. 1972. Preliminary studies of the interactions between feeding vampire bats, *Desmodus rotundus*, under natural and laboratory conditions. *Extrait de Mammalia* 36(2): 241–46.

Schmidt, U.; Greenhall, A. M.; and Lopez-Forment, W. 1970. Vampire bat control in Mexico. *Bijdragen tot de Dierkunde* 40(1): 74–76.

——. 1971. Oekologische Untersuchungen der Vampirfledermäuse (*Desmodus rotundus*) im Staate Puebla, Mexiko. *Z. f. Säugetierkunde* 36(6): 360–70.

Schmidt, U., and Manske, U. 1973. Die Jugendentwicklung der Vampirfledermäuse (*Desmodus rotundus*). *Z. f. Säugetierkunde* 38: 14–33.

Schmidt, Uwe, and van de Flierdt, Kathrin. 1973. Innerartliche Aggression bei Vampirfledermäusen (*Desmodus rotundus*) am Futterplatz. *Z. Tierpsychol.* 32: 139–46.

Schoener, Thomas W. 1971. Theory of feeding strategies. In *Annual Review of Ecology and Systematics*, ed. Richard F. Johnston, P. W. Frank, and C. D. Michener, pp. 369–404. Palo Alto: Annual Reviews.

Schumacher, F. X., and Eschmeyer, R. W. 1943. The estimation of fish populations in lakes and ponds. *J. Tenn. Acad. Sci.* 18: 228–34.

Shannon, R. C. 1928. Zoophilous moths. *Science* 68(1767): 461–62.

Springer, S. 1960. Natural history of the sandbar shark *Eulamia milberti*. *U.S. Fish Wildl. Serv. Fish. Bull.* 178(61): 1–38.

Steele, J. H. 1966. International aspects of veterinary medicine and its relation to health, nutrition and human welfare. *Milit. Med.* 131: 765–78.

Sugay, W., and Nilsson, M. R. 1966. Isolamento de virus da raiva de morcegos hematofogos do estado de São Paulo, Brasil. *Bol. Of. Sanit. Panam.* 60(4): 310–15.

Suthers, Roderick A. 1970. Vision, olfaction and taste. In *Biology of Bats*, ed. William A. Wimsatt, vol. 2., pp. 265–310. 3 vols. New York: Academic Press.

Suthers, R. A.; Chase, J.; and Bradford, B. 1969. Visual form discrimination by echolocating bats. *Biol. Bull.* 137: 535–46.

Thompson, R. Dan; Mitchell, G. Clay; and Burns, Richard J. 1972. Vampire bat control by systemic treatment of livestock with an anticoagulant. *Science* 177: 806–8.

Torres, S., and de Queiroz Lima, E. 1936. A raiva e os morcegos hematophagos. *Rev. Dep. Nac. Prod. Anim.* (*Rio de Janeiro*) 3: 165–74.

Trapido, Harold. 1946. Observations on the vampire bat with special reference to longevity in captivity. *J. Mammal.* 27(3): 217–19.

Turner, Dennis; Shaughnessy, Anna; and Gould, Edwin. 1972. Individual recognition between mother and infant bats (*Myotis*). In *Animal orientation and navigation, a symposium*, ed. Sidney R. Galler, K. Schmidt-Koenig, G. Jacobs, and R. Belleville, pp. 365–71. Washington, D.C.: National Aeronautics and Space Administration.

Villa-R., B. 1966. *Los murcielagos de Mexico.* Mexico City: Instituto de Biologia, Universidad National Autonoma de Mexico.

——. 1968. Ethology and ecology of vampire bats. *International Union for the Conservation of Nature Publications, New Series* (Morges), 13: 104–10.

Vincent, F. 1963. Acoustic signals for autoinformation or echolocation. In *Acoustic behavior of animals*, ed. R. G. Bushnel, pp. 183–227. Amsterdam: Elsevier.

Voisin, A. 1959. *Grass productivity*, trans. C. T. M. Herriot. New York: Philosophical Library.

Whittow, G. C. 1962. The significance of the extremities of the ox (*Bos taurus*) in thermoregulation. *J. Agric. Sci.* 58: 109–20.

Williams, H. E. 1960. Bat-transmitted paralytic rabies in Trinidad. *Can. Vet. J.* 1: 20–24, 45–50.

Wimsatt, William A. 1959. Attempted "cannibalism" among captive vampire bats. *J. Mammal.* 40(3): 439–40.

——. 1969. Transient behavior, nocturnal activity patterns and feeding efficiency of vampire bats (*Desmodus rotundus*) under natural conditions. *J. Mammal.* 50(2): 233–44.

Wimsatt, William A., and Guerriere, Anthony. 1962. Observations on the feeding capacities and excretory functions of captive vampire bats. *J. Mammal.* 43(1): 17–27.

Wimsatt, William A., and Trapido, Harold. 1952. Reproduction and the female reproductive cycle in the tropical American vampire bat, *Desmodus rotundus rotundus. Am. J. Anat.* 91: 415–45.

World Health Organization Expert Committee on Rabies. 1966. *Fifth Report.* WHO Tech. Rpt. No. 321. Geneva.

Young, Allen M. 1971. Foraging of vampire bats (*Desmodus rotundus*) in Atlantic wet lowland Costa Rica. *Rev. Biol. Trop.* 18(1, 2): 73–88.

INDEX

THE JOHNS HOPKINS UNIVERSITY PRESS
This book was composed in Times Roman text and Univers Bold display
type by Jones Composition Company from a design by Susan Bishop.
It was printed by Collins Lithographing and Printing Co., Inc., on
60-lb. Clear Spring Book Offset paper and bound in Holliston Roxite
cloth by Haddon Bindery, Inc.

Library of Congress Cataloging in Publication Data

Turner, Dennis C   1948–
    The vampire bat.

    Bibliography: pp. 131–39
    1. Vampire bats. 2. Mammals—Costa Rica—Guanacaste (Province)
I. Title.
QL737.C52T87   599′.4   74-24396
ISBN 0-8018-1680-7